Public and Private Participation in the Water and Wastewater Sector

Water Law and Policy Series

The role of law in the development and management of the world's water resources is often overlooked by water resource experts. This series aims to make "water law" – international, national and transnational – more accessible to the global water community.

While focusing primarily on the legal component of water resources management, the series will integrate other water-related disciplines in order to ensure a truly comprehensive and interdisciplinary approach to the problem.

SERIES EDITORS

Patricia Wouters
International Water Law Research Institute, University of Dundee, UK.

Sergei Vinogradov
Centre for Energy, Petroleum and Mineral Law and Policy,
University of Dundee, UK.

Other titles in the series include:

A New Legal Framework for Managing the World's Shared Groundwaters: *A Case-study from the Middle East*

Hydrology and Water Law – Bridging the Gap: *A Case-study of HELP Basins*

Water Rights: *An International and Comparative Study*

Public and Private Participation in the Water and Wastewater Sector:
Developing Sustainable Legal Mechanisms

X Cledan Mandri-Perrott

Published by   IWA Publishing
                       Republic - Export Building
                       1 Clove Crescent
                       London E14 2BA, UK
                       Telephone: +44 (0)20 7654 5500
                       Fax: +44 (0)20 7654 5555
                       Email: publications@iwap.co.uk
                       Web: www.iwaponline.com

First published 2009
© 2009 IWA Publishing

Apart from any fair dealing for the purposes of research or private study, or criticism or review, as permitted under the UK Copyright, Designs and Patents Act (1998), no part of this publication may be reproduced, stored or transmitted in any form or by any means, without the prior permission in writing of the publisher, or, in the case of photographic reproduction, in accordance with the terms of licences issued by the Copyright Licensing Agency in the UK, or in accordance with the terms of licenses issued by the appropriate reproduction rights organization outside the UK. Enquiries concerning reproduction outside the terms stated here should be sent to IWA Publishing at the address printed above.

The publisher makes no representation, express or implied, with regard to the accuracy of the information contained in this book and cannot accept any legal responsibility or liability for errors or omissions that may be made.

Disclaimer
The information provided and the opinions given in this publication are not necessarily those of IWA or of the authors, and should not be acted upon without independent consideration and professional advice. IWA and the Author will not accept responsibility for any loss or damage suffered by any person acting or refraining from acting upon any material contained in this publication.

British Library Cataloguing in Publication Data
A CIP catalogue record for this book is available from the British Library

Library of Congress Cataloging-in-Publication Data
A catalog record for this book is available from the Library of Congress

ISBN: 9781789065299

Contents

Table of Abbreviations ix
Acknowledgements xi
About the Author xii
Abstract xiii

1   INTRODUCTION 1

2   THE IMPACT OF EUROPEAN UNION ACCESSION
     ON WATER UTILITIES 9
     2.1   EU and its implications for the water sector of
           Accession Countries 9
     2.2   Context of European integration 10
     2.3   European Enlargement and its Accession Process 13
           2.3.1   General Accession Criteria 16
           2.3.2   Drivers for implementing Accession Criteria 17

3   PUBLIC PRIVATE PARTNERSHIP AS AN INSTRUMENT
     TO EU ACCESSION 21
     3.1   Public Private Partnerships: A possible solution? 21
     3.2   Considerations for PPP in the provision of water
           and wastewater services 23
     3.3   Project Finance, PPP Options and Implications 25
           3.3.1   Service Contracts 27
           3.3.2   Management Contracts 28
           3.3.3   Lease Contracts 29

	3.3.4	Concession Contracts 31
	3.3.5	BOOT Contracts 32
	3.3.6	Joint Ownership 34
	3.3.7	Outright Sale/Divestiture 35

4 A SUMMARY ON ECONOMIC REGULATION AND ITS IMPLICATIONS ON PPPS IN THE EU 37

4.1 Basic Principles of Regulation 37
4.2 Regulation of Public and Private service providers 40
4.3 Regulation, National Policy and the EU legislative context 41
4.4 Developing a Regulatory framework for effective PPP 43
- 4.4.1 Predictability 45
- 4.4.2 Adequate alignment of the PPP contract with national law 45
- 4.4.3 A well defined PPP contract 47
- 4.4.4 A degree of flexibility to deal with long contract periods 47
- 4.4.5 Independence, accountability and transparency of regulation 48

5 EUROPEAN LEGAL FRAMEWORK AND ITS IMPLICATIONS ON PPP 49

5.1 EU's Approach to PPP 49
5.2 Relevance of EU Law in the water sector 51
- 5.2.1 EU Utilities Directive 52
- 5.2.2 Implications of public involvement & ownership under EU 54
- 5.2.3 Implications of State Aid 55
- 5.2.4 EU Competition Law 55
- 5.2.5 Other considerations related to EU Competition Law 57

5.3 EU Legal Position on PPP 58
5.4 EU Contracting options for PPP 59
- 5.4.1 Implications of PPP, national law and concessions 60
- 5.4.2 Purely Contractual PPP 62
- 5.4.3 PPPs as concessions 62
- 5.4.4 Institutionalised PPPs (IPPPs) 63

	5.5	Water Framework Directive – Background & Development	64
		5.5.1 Brief analysis of the WFD	66
		5.5.2 Summary of environmental objectives of the WFD	67
		5.5.3 Summary of Managerial objectives of the WFD	68
		5.5.4 Framework for analysis of WFD in Water & Wastewater Utilities	70

6 FRAMEWORK TO ACHIEVE CONSISTENCY BETWEEN A PPP STRUCTURE AND EU LAW ... 73

- 6.1 Rationale behind the Propopsed Framework to achieve consistency between a PPP Structure and EU law ... 73
- 6.2 Overview of the Proposed Framework for PPP compliance with EU law ... 76
  - 6.2.1 Framework 1: EU Competition Law ... 76
  - 6.2.2 Framework 2: Regulatory Provisions ... 77
  - 6.2.3 Framework 3: Water Framework Directive ... 77

7 THE SOFIA, BULGARIA CONCESSION CONTRACT ... 79

- 7.1 Bulgaria – Historical Background to PPP ... 79
- 7.2 Private Sector Participation in Sofia ... 81
- 7.3 Overview of PPP Development & Process ... 82
- 7.4 Application of Framework 1: EU Competition Law ... 84
- 7.5 Application of Framework 2: Regulatory Provisions ... 89
- 7.6 Application of Framework 3: Water Framework Directive ... 92
- 7.7 Bulgaria Sofia Summary of findings ... 96

8 TALLINN, ESTONIA PARTIAL DIVESTITURE/SALE CONTRACT ... 99

- 8.1 Estonia – Historical background to PPP ... 99
- 8.2 Private Sector Participation in Tallinn ... 101
- 8.3 Overview of PPP Development & Process ... 102
- 8.4 Application of Framework 1: EU Competition Law ... 105
- 8.5 Application of Framework 2: Regulatory Provisions ... 111

	8.6	Application of Framework 3: Water Framework Directive	114
	8.7	Estonia Tallinn Summary of findings	118
9	**CONCLUSIONS**		**121**
ANNEX 1:	Framework 1: EU Competition Law		127
ANNEX 2:	Framework 2: Regulatory framework for Analysis		129
ANNEX 3:	Framework 3: Water Framework Directive (WFD)		131
ANNEX 4:	Diagrammatic Representation of Test		135

BIBLIOGRAPHY			**137**
A.	**Primary Sources**		137
	A.1	Treaties	137
	A.2	EU Directives	137
	A.3	Other EU Measures	138
	A.4	Legal Documentation of PPP processes (Sofia and Tallinn)	138
B.	**Secondary Sources**		139
	B.1	Books	139
	B.2	Articles and presentations	140
	B.3	Reports and Periodicals	141
	B.4	Interviews	142
	B.5	EU Cases and European Court of Justice decisions	143
	B.6	Internet Resources	143

Table of Abbreviations

ASTV	AS Tallinna Vesi – water and wastewater operating company for the City of Tallinn
BAT	Best available technique
BATNEEC	Best available technique not entailing excessive costs
bn	Billion
BOOT	Build Own Operate and Transfer
CMU	Concession Monitoring Unit
CoT	City of Tallinn
EBRD	European Bank for Reconstruction and Development
EC	European Community
ECJ	European Court of Justice
ECSC	European Coal and Steel Community
ECR	European Court of Justice Reports
ed	editor
EEA	European Environment Agency
EU	European Union
FT	Financial Times
Fn	Footnote
GDP	Gross Domestic Product
HELCOM	Helsinki Convention on Baltic Environmental Protection
ISPA	Instrument for Structural Policy for Accession Countries
km	kilometre
lt	litres

© 2009 IWA Publishing. *Public and Private Participation in the Water and Wastewater Sector: Developing Sustainable Legal Mechanisms*, by X Cledan Mandri-Perrott. ISBN: 9781789065299. Published by IWA Publishing, London, UK.

m	miles
mm	millimetre
MMU	Mandate Monitoring Unit
MoS	Municipality of Sofia
MRDPW	Bulgarian Ministry for Regional Development and Public Works
NGO	Non Governmental Organisation
No	Number
PO	Private Operator
PO/SI	Private Operator / Strategic Investor
RBD	River Basin District
RWA	Regional Water Authorities
Sec	Section
SHA	Shareholders Agreement
TACIS	Technical Assistance for the Commonwealth of Independent States
TEC	Treaty of the European Community
WA	Water Act
Water Services	The terms *water services* and *water and wastewater services* are considered as interchangeable throughout this book
WB	World Bank
WFD	Water Framework Directive
v	versus
ViK	Vodokanal – Water and wastewater company

Acknowledgements

I dedicate this work to my wife Yvonne without whom none of this would have been possible: to her unrelenting support and love, and for always believing in me, and to our son Dominique and daughter Stephanie who have brought such sunshine to our lives. A debt of gratitude to my Mother for her encouragement throughout.

About the Author

X Cledan Mandri-Perrott

Cledan combines practical water operational knowledge with policy and private sector participation expertise. His practical operational experience comes from working for an international water operator and as Director of a local joint venture company for a three-year management contract in Trinidad and Tobago. He has been involved in the development of infrastructure Public Private Partnerships structures in various countries, advising both Governments and Private companies. He has strong regulatory background and has planned, designed and implemented regulatory and policy frameworks. Furthermore, he has assisted a number of water utilities in evaluating the potential benefits of PPPs. He has recently gained a reputation for dealing with PPP issues that are in distress either through dispute resolution and mediation with regulators or the re-design of failed PPP structures. Notably, he was the lead adviser for the PPP transaction in Estonia. In Bulgaria he was engaged by the City of Sofia to assist in the setting up of the Concession Monitoring Unit, one of the conditions precedent to effectiveness of the Sofia Concession. This book draws on the experiences of these two transactions, but also benefits from his involvement in other PPP water deals including Argentina, Chile, Jordan, Nepal, Nicaragua, Russia, Tanzania, and Uzbekistan amongst others. Cledan is a Civil Engineer by training, with a Bachelor in Civil Engineering from Surrey University, and has a Masters degree in Property, Business and Management, and an LLM from the University of Dundee, UNESCO Centre for Water Law, Policy and Science.

© 2009 IWA Publishing. *Public and Private Participation in the Water and Wastewater Sector: Developing Sustainable Legal Mechanisms*, by X Cledan Mandri-Perrott. ISBN: 9781789065299. Published by IWA Publishing, London, UK.

Abstract

Europe has a wide range of languages, religions, eternal history of war, constant changes in national boundaries and a host of idiosyncrasies hardly conducive to unification. However, the successful growth of the European Union from its original membership to its actual 27 members, demonstrates a trend towards economic unification of Europe.

This book discusses viable international legal structures in the management, distribution and treatment of one of the most vital, if not *the* most vital, element for the sustenance of humankind – water. In so doing, there is a brief overview of the present realities of European integration, the political and legal aspects involved in the water sector and two cases in which a viable solution was reached and which are the basis of this book.

The need for financing infrastructure services and in particular the water and wastewater sector is a challenge being faced by a number of governments seeking accession to the European Union. Moreover, Public authorities at all levels, including Member Countries are increasingly interested in developing specific arrangements for cooperation with the private sector. The Public Private Partnership approach has a track record in establishing successful cooperation arrangements, and is the subject of detailed review here, in order to show the issues involved. Despite the increasing use of PPP in both Member and Accession Countries, Community law does not provide for specific legislation on PPP. However, any delegation of responsibility may be the subject of Community law. Accordingly, the form of delegation through a PPP instrument needs to be examined against the basic principles of EU law and the framework it establishes. EU law is silent on whether a public authority should enter into a

© 2009 IWA Publishing. *Public and Private Participation in the Water and Wastewater Sector: Developing Sustainable Legal Mechanisms*, by X Cledan Mandri-Perrott. ISBN: 9781789065299. Published by IWA Publishing, London, UK.

PPP. The decision to delegate an economic activity to a third party is squarely in the hands of the public authority.

In the absence of a uniform legal framework for PPPs at a European level and in view of the heterogeneity of national laws, the challenge is to ensure transparency and effective competition between market participants. Accordingly, the book examines the general principles of EU law in terms of competition and procurement and how other directives have an impact on PPP. It then assesses the specific rules applicable to PPP in the EU context, and specifically their implications in designing water and wastewater PPPs.

PPPs may be one of the various instruments available for governments towards meeting some of the criteria for accession to the EU. Particularly aspiring countries need to comply with environmental standards which require members to adhere to a common approach to the environment. This includes the crucial provision of water and wastewater services and its management by member countries and accordingly the EU's Water Framework Directive is examined in this context. EU's procurement principles (under the Utilities Directive and to some extent Competition law), Water Framework Directive, and principles of economic Regulation are used to form the basis of the framework for analysis of two case studies in Bulgaria and Estonia. This book provides a practical methodology and framework to help better design PPP structures against three main criteria:

Criteria 1 relates to EU Competition Law and seeks to determine, under the given form of PPP structure developed what is the extent of public involvement, relevant market of the service provider, potential for abuse of a dominant position and how competition principles are dealt with.

Criteria 2 looks at whether basics principles of economic regulation are adequately incorporated into the PPP structure. In particular it seeks to understand whether the proposed contractual and institutional structure provides for a good definition of roles and responsibilities, independence, accountability and transparency between the parties.

Criteria 3 deals specifically at the EU's Water framework Directive and how the proposed PPP structures seek to achieve the different WFD principles though the PPP contract obligations.

The question of whether or not Accession Countries can use the private sector as an effective tool to achieve compliance with Community law and its directives and thus assist in the accession process is answered.

The book concludes with a detailed review of the two case studies (the City of Sofia, Bulgaria and the City of Tallinn, Estonia) that shows how the Public Private Partnership structure chosen provides a sound legal basis and a viable way to achieve compliance with Community law and the WFD, thus assisting the accession process of each country.

This book is principally aimed at supporting municipal, provincial, and central governments and other policy makers seeking to improve water and wastewater services. Although the examples relate to EU Accession countries, the principles addressed in this book may be used outside the European context.

1
Introduction

The past few decades have seen a growing awareness of the planet's fragile ecological balance. Water resources are being stretched to their limits and are putting pressure on urban centres worldwide. Nevertheless, there is a growing consciousness of the need for peoples to unite to achieve policies that will assure the protection of this precious resource and the environment within their economic framework. This herculean task of unifying peoples in different parts of the world might be considered viable in say, Latin America, where there is a vast expanse of land with a common language, and a common imprint of five hundred years of Spanish. However, perhaps the most striking example is the unification of Europe, with its wide range of languages, religions, and eternal history of war, constant changes in national boundaries and a host of idiosyncrasies – hardly conducive to unification. Recently, the European Union has successfully grown from its original 6 members to 27 members today. This unification is a remarkable feat and is fast becoming a reality.

© 2009 IWA Publishing. *Public and Private Participation in the Water and Wastewater Sector: Developing Sustainable Legal Mechanisms*, by X Cledan Mandri-Perrott. ISBN: 9781789065299. Published by IWA Publishing, London, UK.

It is in the context of European unification that this book will be discussing viable international legal structures in the management, distribution and treatment of one of the most vital, if not *the* most vital, element for the sustenance of humankind – water.

The realities of European integration
The fact that the European Union has successfully grown from 6 to 27 members is remarkable.[1] However, the requirements for entry are stringent[2] and demand careful consideration, both on behalf of the EU and particularly for those countries aspiring membership – the Accession Countries.[3] Environmental standards are being set across the European Union[4] and Accession Countries have to comply with these. Furthermore the EU requires its members to adhere to a common approach to the environment, which crucially includes the provision of water and wastewater services and its management by member countries. Accordingly, countries' public authorities are increasingly looking at partnering with the private sector in the form of Public Private Partnerships (PPP) as an instrument that can be used to obtain improved infrastructure. These partnership structures must fully conform to the requirements of the EU law and the principles of transparency, fairness, etc as well as to European Water Framework Directive (WFD).[5]

Can the public and the private sector collaborate or partner to assist Accession Countries?
The importance of the public and the private sector partnering through Public Private Partnerships (PPPs) will be discussed, as well as how the approach seeks to find pragmatic and sustainable solutions to the key problem of reconciling the need to meet accession criteria with the massive funding requirements for water utility[6] infrastructure. Since the private sector plays an important role in accessing funds and bringing management, technical and commercial expertise to existing water utilities, PPP structures are seen as a possible mechanism

---

[1] *See* more detailed discussion on this point, *infra*, at 11–13.
[2] European Council, Madrid Proceedings December 1995, taken from http://europa.eu.int
[3] UNEP, *The Use of Economic Instruments for Sustainable Development*, Meeting on March 2001. http://www.unep.ch/etu/etp/events/recent/ber_econinst.pdf, at 1: Accession Countries have historically been characterised by low investment, low service standards and quality of government subsidies and cross-subsidies, poor utility performance and low cost recovery.
[4] *See* European Council, Madrid Proceedings December 1995, *supra* note 2.
[5] The Water Framework Directive (WFD) is a part of European Union legislation which states the principles required of each country to encourage efficient and effective water protection.
[6] *'Water Utility'* is used to indicate providers of water and/or wastewater services.

to ensure compliance[7] with accession criteria and with the WFD's[8] basic principles. At the same time, European legislation at the Community level does not explicitly provide for 'public private partnerships' or PPPs. There are no strict definitions for PPP within Community legislation although some countries or member states (including accession countries) have created specific tools to coordinate and promote PPPs.

Despite the fact that EU does not explicitly allow for detailed PPP legislation, partnering between the public and private sector involves developing adequate legal frameworks that deliver the services and standards required, yet at the same time are flexible, creative, manageable and sustainable. Sustainability itself will depend – amongst other things – upon an adequate balance of risk between the parties. Given these constraints, this book, from the perspective of the water and wastewater sector, looks at providing sector practitioners in Accession Countries with a series of practical tools to use effectively the concept of *'partnering'* with the private sector to address some of the current limitations of the public sector particularly with regard to improving management and financing capacity. In addition the book provides a link with the EU legislative context by assessing whether the structures that aim to introduce the private sector have a sound legal basis and can help to achieve compliance with Community law and directives, and consequently to assist the accession process.

Data presented in this book has been collected through an examination of the historical evolution of the EU, its existing laws, and the EU's approach to the provision of water and wastewater services, as well as the structures available to the private sector in the provision of water services. Given that the EU's position on PPPs has significantly evolved in recent years, the latest 'official' stance is presented in this book, but the reader is advised to check current legislation on this matter. A review of literature on water issues from various disciplines including legal, political, economic and social development has been included. Existing contractual documents have been used as the basis for analysis in the two case studies.

---

[7] Economic instruments, together with adequate monitoring and reporting tools are essential, not only to achieve the WFD's environmental objectives but also to ensure that there are viable channels through which the private sector may be introduced.

[8] Directive 2000/60/EC of the European Parliament and of the Council of 23 October 2000 establishing a framework for Community action in water policies, 2000 O.J. (L327) 1, [hereinafter, the Water Framework Directive (WFD)].

[9] Idelovitch, E., and Ringskog, K., Private Sector Participation in Water Supply and Sanitation in Latin America. (Washington, DC: World Bank, 1995), at 10: The balance of risk between the public and private sectors is essential for successful PPP structures.

Chapter 2 provides a short overview of the historical context in which European integration developed and how expansion has evolved and is currently being managed. The legal and institutional enlargement structure used to integrate Countries into the EU will also be discussed.[10]

This book argues that in a number of cases the introduction of the private sector through the creation of a Public Private Partnership (PPP) may be used effectively as an instrument to mitigate the public sector's limitations in the provision of services to required standards, the mobilisation of financing (both public and private) and thus assists Accession Countries meet accession criteria. Chapter 3 examines the motivation for involving the Private Sector, looks at the basic principles of project finance and provides an overview of the different options and commensurate risk allocation of PPP structures and techniques available. Some of the benefits of PPP involvement are also discussed in this chapter.

In most countries, specific regions are served with water and wastewater services provided by a single entity. Structuring utilities as 'natural monopolies'[11] helps to achieve universal accessibility and affordability and concentrate investment into one single infrastructure network. Inevitably, citizens have no choice but to receive service from this sole provider. The resulting combination of vital services and monopoly pricing power requires some form of 'regulation'. In Chapter 4, we examine the basic principles of regulation but from an economic perspective in order to ensure that services are provided in an effective way and at an affordable price. So whilst monopolies may represent the best way to achieve universal access at the lowest investment cost, on the other hand, if left unchecked would not set price or supply services in an efficient manner. Reconciling these conflicting interests requires regulation to allow for efficient investment while also compensating for the absence of market forces. Under this chapter we develop the first framework for analysis of the two case studies presented in this book. Practical understanding of the significance of the regulatory issues is achieved by identifying a series of regulatory design considerations whose inclusion is a necessary means for developing a framework for effective and sustainable PPP structures in Accession Countries.

Chapter 5 is concerned with EU legislative framework. We examine the impact European Law principles have on the provision of water and wastewater services. In the water sector, water services tend to be offered by a single entity

---

[10] *See* more detailed discussion on this point and specifically *acquis communautaire*, *infra*, at 14.

[11] Usually there is no more than one provider of water and water services in a given region, therefore giving rise to a 'monopoly situation'. However, there are cities in which there may be two distinct service providers. In either case there is not an open economic condition in the provision of services and as such, there is a risk of 'monopoly'.

in a monopoly fashion. Given this monopolistic situation and the fact that the services provided have major economic and social impacts that put the service provider at an undue advantage with respect to the consumer/costumer, it is important that abuses do not occur. EU Competition law is concerned with this situation. Accordingly, it is suggested that a series of tests may be developed to assess any abuse of a dominant position, to determine the extent of public involvement, to identify which would be the relevant market(s), to see how much assistance has been given by the State and to devise a regulatory mechanism to support the Competition principles. This forms the second framework for analysis of the book's case studies.

Notably, within Community law there are no prescriptions at law for PPP relationships nor, in fact, whether a public authority should enter into a PPP. The decision to provide an economic activity is squarely in the hands of the public authority. If it wishes to delegate a part or the whole of this activity to a third party it is the prerogative of the public authority. Accordingly, in this chapter we review the impact Community law may have on developing PPP structures. This will be used later on as a framework for analysis for the case studies developed in this book. In the absence of a uniform legal framework for PPPs at European level and in view of the heterogeneity of national laws the question arises, whether uniform provisions are needed to safeguard transparency and effective competition between market participants.

Furthermore, in what regards the water and wastewater sector, it will be shown how the EU sets the required result in a *'directive'*, but that it is up to each state to create its own legislation to achieve that desired outcome. In this general context, the particular directive affecting the water and wastewater sector – the Water Framework Directive (WFD) – is examined, its main principles, objectives and evolution explored. Particular emphasis has been placed on the principles of water quality, management by river basin, monitoring and reporting requirements and management of the water and wastewater process through the principle of cost recovery. This forms the first element in the framework for analysis of this book.

Chapter 6 brings together the three elements of the multi-criteria framework based on three key factors driving PPP in the water and wastewater sector in Accession Countries. These are the European Competition Law, Regulation and

---

[12] Usually there is no more than one provider of water and water services in a given region, therefore giving rise to a 'monopoly situation'. However, there are cities in which there may be two distinct service providers. In either case there is not an open economic condition in the provision of services and as such, there is a risk of 'monopoly'.

the Water Framework Directive. These will be used as the basis for examining the two relevant case studies in Sofia-Bulgaria and Tallinn-Estonia.

Application of the methodological framework to the case studies
Having discussed the evolution of the EU, its law, and the available options to provide sustainable PPP structures, we provide a detailed analysis of two case studies and respective PPP structures in Chapters 7 and 8: Sofia-Bulgaria and Tallinn-Estonia.[13] The multi-criteria framework (EU Competition Law, Regulation and WFD) is then systematically applied to each country's chosen form of PPP. By combining both theoretical and practical data, an assessment is made of the effectiveness with which each country deals with accession criteria (WFD), the legality of the chosen structure, existing contractual terms and it is overall.

Conclusions
The book concludes that for both the City of Sofia, Bulgaria and the City of Tallinn, Estonia, the PPP structure chosen provides a sound legal basis and a viable way to achieve compliance with Community law, the basic principles of economic sector regulation and the WFD, thus assisting the accession process of each country.[14] This investigation places significant emphasis on the regulatory aspects of the provision of water services, and as such, examines them in detail. Both case studies used are recent contracts still in their infancy, but the book suggests through the methodological framework used to analyse each contract that both contracts have been effective and successful in helping Estonia and Bulgaria meet the accession criteria. However, full effectiveness cannot be really measured until these contracts are more mature.

Notwithstanding the above, the book provides a mechanism for analysing aspects to be considered when designing PPPs, or for assessing the suitability and viability of the PPP alternative.

The general conclusion drawn from these two examples is that PPP contracts should be flexible, creative, manageable and sustainable if they are to effectively act as a vehicle to achieve EU accession criteria. Furthermore, among a number of lessons to be learnt and applied to other Accession Countries, is that each case is unique and its idiosyncrasies should be considered carefully.

---

[13] These examples have been selected as each represent differing characteristics in population, geographical location and WFD relevance e.g. the City of Sofia, Bulgaria is land locked, and the City of Tallinn, Estonia, discharges directly into the Baltic Sea. In addition, each country represents a different stage in the Accession process. Estonia was ultimately accepted to the EU on 1 May 2004, and Bulgaria joined on 1 January 2007.

[14] When this research for this book started Estonia was not yet part of the EU. It reached accession in 1 May 2004. Bulgaria has also subsequently joined the EU on 1 January 2007.

Introduction

Areas for future research

As mentioned above, it is too early to generalize about the effectiveness of PPP structures. The framework used in this book can be expanded to help design structures and explore their possibilities and limitations in the drive towards compliance with the WFD and eventual accession to the EU. It is outside the scope of this book to analyse the ways in which the limitations can be addressed. For example; How could the public sector restructure to create the same advantages as one of the PPP structures examined? How to balance the profit motivation of the private sector with government's obligation to provide services to the poor? How are subsidies managed? This investigation does not explore how other EU Directives affect the provision of water services. Similarly, the book does not delve deeply into the regulatory aspects of setting tariffs.

2
The impact of European Union accession on water utilities

2.1 EU AND ITS IMPLICATIONS FOR THE WATER SECTOR OF ACCESSION COUNTRIES

This chapter provides a brief history of the development of the European Union and the basic principles that underpin such a union. The evolution of the European Union will be discussed and in particular the manner in which enlargement of the EU is being developed and managed. The principal focus is to provide a framework by which Accession Countries are integrated into the enlargement process of the Union. Accordingly, the European Commission has developed various mechanisms in support of the Accession Countries. The accession process will be examined and the reasons behind the implementation of the criteria, developed by the European Union, that have to be met by Accession Countries.

© 2009 IWA Publishing. *Public and Private Participation in the Water and Wastewater Sector: Developing Sustainable Legal Mechanisms*, by X Cledan Mandri-Perrott. ISBN: 9781789065299. Published by IWA Publishing, London, UK.

2.2 CONTEXT OF EUROPEAN INTEGRATION

In order to fully understand the impact of European integration we have to be aware of the context in which it occurs.[15] Understanding this partly hinges on the historical, political, cultural and economic contexts that have led to its formation. Moreover, it is important to understand how events such as the collapse of the Soviet Union and the liberalisation of Eastern and Central Europe (including Eastern Germany) have been instrumental in driving European integration.[16] However, apart from these macro factors, there were other important reasons, not the least being to prevent war through economic and military integration.[17&18] These in

---

[15] Vincenzi, C., Law of the European Community, 5, Financial Times Pitman Publishing, England (1999). The recognition for an all-inclusive Europe has its origins firmly placed in historical conflict. At the end of the First World War, the Treaty of Versailles (1919) recognised the new geographical boundaries resulting from the collapse of the Austro-Hungarian and Ottoman empires. The treaty's architects (Britain, France and Italy) imposed heavy reparations on Germany. Suffering from hyperinflation and economic collapse, the new Weimer republic was unable to pay. Moreover, Britain, France and Italy were also heavily affected and attempts to protect national economies by tariff barriers were largely unsuccessful. The stock market crash of 1929 nullified the economy and pushed unemployment up to 40%. A direct result of this instability led to the rise of the Nazi Party and the outbreak of the Second World War. It was not until preparations for the Second World War that the economies of these countries were "jump started" once again.

[16] A Survey of European Integration, The Economist, May 19th 2001, at 3: After more than 40 years the iron curtain divided the continent of Europe between a prosperous and free west and an impoverished and oppressed east. When Communism collapsed and the iron curtain came down in 1989, the European Union (formerly known as the European Economic Community and comprised of only six members) pledged to embrace the countries of the east by admitting them to it. This, it was hoped would spread the peace, stability and prosperity enjoyed by the West to the east and "re-unify" the continent. More than a decade later, the Union looks likely at last to make good its promise

[17] Hobsbawm, E., Age of Extremes, The Short Twentieth Century, (1994), Michael Joseph, Chapters 2 & 3, at 59: France, who had fought with Germany three devastating wars between 1870 and 1945, was painfully aware of the need to avert war. After the First World War the linked defence treaties and the League of Nations had been unable to deter war. The French in particular understood the importance of binding Germany's coal and steel industry into a new political and economic alliance. "For France the main point of European integration was to prevent Germany from again dominating Europe. As a senior French official working for the EU puts it: "France has still not got over psychological shock of German unification. Preventing the rise of German power has been a French national goal since the times of Richelieu." After the Second World War and recognising the need for mutual interdependence by the European States and as such, the resurgence of ideas about European political unity, Winston Churchill expressed these views in the Zurich Conference of September 1946. In his speech he proposed the construction of a "kind of United States of Europe" based on a Franco-German partnership".

[18] *See* The Economist, *supra* note 16 at 14: At the same time, the Soviet Union now occupied the whole of Central and Eastern Europe including East Germany. The democratic countries of Western Europe and the United States were concerned about a perceived expansionist Soviet

turn resulted in a recognition that European economies are inextricably linked and could not be successfully managed independently.[19] In order to manage these joint economies, it became evident that there was a clear need to establish a legal instrument[20] to be recognised by all Member Countries. This higher authority would be empowered to develop policies which, although they might temporarily affect an individual country adversely, had the greater good of all countries at heart.[21] There was also recognition that a European union of various countries could only work together[22] in an environment characterised by unrestricted[23] and fair[24]

Union. Thus, in 1949 they met to form the North Atlantic Treaty Organisation (NATO) with the United States of America at the core of the organisation. During this period the Marshall plan provided millions of dollars in aid for various European states that marked the resurgence of the European economy.

[19] *See* Hobsbawm, *supra* note, 17 at 28: The experience of the inter-war years clarified beyond doubt that it was no longer possible for the states of Western Europe to operate their national economies without regard to the effect they may have on their immediate neighbours.

[20] The European Coal and Steel Community (ECSC) Treaty 1951 (Preamble). In 1950, the European Coal and Steel Community (ECSC) Treaty signed in April 1951 by Germany, Belgium, France, Italy, the Netherlands and Luxembourg made it clear that the long-term aims were much more ambitious than the control of coal and steel production. The participants were resolved to create, by establishing an economic community, the basis for a broader community among peoples long divided by bloody conflicts.

[21] Hartley, T.C., The Foundations of the European Community, (1994) Clarendon Press, London, at 45: However, the most striking fact about the ECSC Treaty was that it established a Community that had legal personality by instituting a High Authority that was to be responsible for the coal and steel policies in the Members Countries, and with granting the High Authority the power to make decisions that may affect a country's economic agents regardless of the wishes of its particular government.

[22] With the Treaty of Rome the European Economic Community (EEC) was born. The basic premise of the EEC was to allow a free trade area in goods, capital, labour and services amongst the Member Countries (MC's). Central to this premise was the abolition of all customs and duties as well as quantitative restrictions in trade between the MC's by the establishment of a single customs union.

[23] Pinder, J., European Community: The Building of a Union (1991) Oxford University Press, at 75: Competition was thus promoted between members of the European Economic Community (EEC's) who set itself the target of creating a single market.

[24] Also central in its strategy to promote competition was the legal ability granted by Member Countries to the EEC to legislate on public procurement issues including competition for public contracts in the whole Community area.

internal trade.[25] At the same time, it was acknowledged that social well-being[26] was intertwined with economic development.[27]

Furthermore, as social consciousness grew and Europe became more economically stable, it became evident in the 1980s[28] that there was an increasing emphasis on environmental issues. The history of the European Union can be traced through the various treaties[29] that address all these issues in preparation for its enlargement[30] and management.[31] Leading up to 1965 there were a number of European institutions all acting in a disparate manner. There was a clear need to integrate the European Coal and Steel Community (ECSC), European Economic Community (EEC) and Euratom as a single unit with simplified administrative structures. In this regard, the Merger Treaty was signed on 8 April 1965, the result of which was only one Council, one European Commission and one European

---

[25] European Community Treaty Article 8A (this was added by the Single European Act), at 8: "....creating a Single Market by removing all the remaining legal, technical and physical obstacles to the free movement of goods, persons, capital and services by 1 January 1993."

[26] *See* The Economist *supra* note 16: It was with the European Coal and Steel Community (ECSC) Treaty 1951 that a social agenda first appeared, not only as a national objective but also as a European one. Policies were developed that allowed for training, housing and re-deployment. Yet it was not until the Treaty of Rome in 1957 that trade barriers were fully removed and a common commercial policy developed as a means of ensuring a social dimension.

[27] *See* Treaty of Rome (1957), at 16: To secure the constant improvement of the living and working conditions of their peoples....[and] to ensure their harmonious development by reducing differences existing between various regions.

[28] It is interesting to note that the UK did not actually join the European Economic Community (EEC) and forged ahead with a much looser free trade association called the European Free Trade Area (EFTA). By 1961 the UK Government recognised that although trade had significantly increased it lacked the "structure and coherence of the EEC". Finally in 1973 the UK joined the EEC.

[29] Weigall, D., and Stirck, A., The Origins and Development of the European Community, (1992) Leicester University Press, England.

[30] *"Enlargement"* is used, to describe the current EU expansion process i.e. countries who wish to join the EU. At the time research for this book began, both Estonia and Bulgaria were considered countries aspiring to membership. Estonia was accepted to the EU in May 2004 whereas Bulgaria joined in 2007. A distinction is made with "Enlargement" as defined in the Treaty of Accession signed 22 January 1972 in which the UK, Ireland and Denmark became members of the European Economic Community (EEC) and the further enlargement resulting from Greece's membership in 1981 and Spain and Portugal's inclusion in 1986. The EU's membership of 15 was completed on the 1 January 1995 with the integration of Finland, Austria and Sweden. Its current membership of 25 includes Cyprus, the Czech Republic, Estonia, Hungary, Latvia, Lithuania, Malta, Poland, Slovakia and Slovenia.

[31] Enlargement and Civil Society, Conference Proceedings 1999, http://europa.eu.int/comm/enlargement/docs/confpro1999.doc. In all, 12 countries are currently negotiating accession to the European Union: Bulgaria, Cyprus, the Czech Republic, Estonia, Hungary, Latvia, Lithuania, Malta, Poland, Romania, Slovakia and Slovenia. In addition, Turkey has been recognised as a candidate for membership.

Court of Justice. After this the European Community saw an unprecedented enlargement in its membership and as such, in its sphere of influence that culminated in the signing of the Single European Act (SEA) in February 1986. This treaty sought to allay the concerns of the Member Counties, which were gradually shifting from mass unemployment in the mid 1970's and early 1980's, to a growing concern about the degradation of the natural environment. It can be seen that before the SEA, the response to environmental concerns was at a national level. Clearly with each country developing its own standards this became one of the key motivational drivers to avoid a diffuse and uncoordinated approach to environmental issues. This haphazard approach seriously threatened the growth of a genuinely common market of goods and services in the EEC. Thus, the SEA extended EEC competence to enable it to legislate for the whole area of the Community on environmental matters, including areas of health, safety, consumer protection, public procurement, excise and duties, frontier controls, research and technical development. Although the SEA allowed for greater powers to be conferred to the EEC, it was not until the Treaty of Maastricht on the European Union, signed on 7 February 1992, that an effective and integrated measure to deal with environmental concerns was fully developed. This treaty is very significant as it intended to extend further competencies of the Community by creating two new "pillars" outside the legally binding formal decision-making process of the other Communities (EC, ECSC and Euratom). These comprised of a common foreign and security policy pillar, and co-operation in home affairs and justice. In addition, the Treaty of Rome amended the EEC to create simply the European Union. This marked a legal recognition that activities and competencies of the former Economic Community ranged far beyond its original economic goals.

2.3 EUROPEAN ENLARGEMENT AND ITS ACCESSION PROCESS

Given the factors driving European integration described above, the expansion, or enlargement of Europe has shifted from purely an arrangement of economic convenience to something broader encompassing social, political, cultural and other aspects including human rights and the environment. This broader agenda was ratified in the Treaty of Amsterdam signed in 1997.[32] The Treaty of Amsterdam had two main functions. Firstly, to deal with the administrative aspect related to the EU's ability to function effectively in direct response to its

---

[32] Treaty of Amsterdam (1997), [hereinafter, Treaty of Amsterdam].

14   Public and Private Participation in the Water and Wastewater Sector

expansionist agenda (also called *enlargement*) given that the EU was originally created to cater for only six nations. Secondly, to broaden the objectives of the EU significantly to move away from a narrow economic focus to more of a commitment to environmental and human rights issues.[33&34]

At the Luxembourg summit in December 1997, the European Council decided that the enlargement process should involve a European Conference whereby there would be a multilateral framework and accession process bringing together ten central European countries including Cyprus and Turkey (both of whom had formally launched their application process on 12 March 1998). Formal accession negotiations proceeded on 31 March 1998 with six countries, as recommended by the European Commission: Cyprus, the Czech Republic, Estonia, Hungary, Poland and Slovenia.

In March 1998 the EU formally launched the process that would make enlargement possible.[35] It embraced thirteen applicant countries, which were Bulgaria, Cyprus, the Czech Republic, Estonia, Hungary, Latvia, Lithuania, Malta,[36] Poland, Romania, the Slovak Republic, Slovenia and Turkey.

Country	Accession date	Population	Area (km^2)
Republic of Austria	1 January 1995	8,340,924	83,871
Kingdom of Belgium	25 March 1957	10,666,866	30,528
Republic of Bulgaria	1 January 2007	7,640,238	110,910
Republic of Cyprus	1 May 2004	778,700	9,251
Czech Republic	1 May 2004	10,403,100	78,866
Kingdom of Denmark	1 January 1973	5,482,266	43,094
Republic of Estonia	1 May 2004	1,340,935	45,226
Republic of Finland	1 January 1995	5,312,415	338,145

---

[33] *See* Article 7, Treaty of Amsterdam, *supra* note 32, The Treaty of Amsterdam proclaims that *'it is founded on respect for human rights and the rule of law'*, and that the rights of a Member State can be suspended if the Member State has *'been in serious and persistent breach'* of the obligations respecting civil, political and human rights.

[34] *Id.,* Article 2 describes equality between men and women as one of the EU's principal objectives.

[35] Presidency Conclusions Helsinki European Council, 10 and 11 December 1999 DN: DOC/99/16 Date: 1999-12-13 http://europa.eu.int/rapid/start/cgi/guesten.ksh?p_action.gettxt=gt&doc=DOC/99/16|0|RAPID&lg=EN

[36] *See* Enlargement and Civil Society, *supra note* 31 at 21. Malta, which had 'frozen' its application for membership in 1996 due to various internal political reasons, reactivated it in October 1998.

Country	Accession date	Population	Area (km²)
French Republic	25 March 1957	64,473,140	674,843
Federal Republic of Germany	25 March 1957	82,218,000	357,050
Hellenic Republic	1 January 1981	11,125,179	131,990
Republic of Hungary	1 May 2004	10,036,000	93,030
Ireland	1 January 1973	4,339,000	70,273
Italian Republic	25 March 1957	59,619,290	301,318
Republic of Latvia	1 May 2004	2,266,000	64,589
Republic of Lithuania	1 May 2004	3,357,873	65,303
Grand Duchy of Luxembourg	25 March 1957	483,800	2,586
Republic of Malta	1 May 2004	407,810	316
Kingdom of the Netherlands	25 March 1957	16,471,968	41,526
Republic of Poland	1 May 2004	38,115,641	312,683
Portuguese Republic	1 January 1986	10,599,095	92,391
Romania	1 January 2007	21,538,000	238,391
Slovak Republic	1 May 2004	5,400,998	49,037
Republic of Slovenia	1 May 2004	2,025,866	20,273
Kingdom of Spain	1 January 1986	46,063,511	506,030
Kingdom of Sweden	1 January 1995	9,208,034	449,964
United Kingdom of Great Britain and Northern Ireland	1 January 1973	61,003,875	244,820
European Union (27 members) total	–	497,456,009	4,456,304

Notes

- Includes the area (3,355 km²) but not the population (264,172 according to 2006 census) of the territory under control of the un recognised Turkish Republic of Northern Cyprus (TRNC). The Government of Cyprus estimates the total population at 867,600 including Turkish Cypriots entitled to EU citizenship but excluding TRNC citizens who immigrated from Turkey after the 1974 invasion.
- According to European Commission document published in January 2008, p.8: *The following are not part of the territory of the European Community: French Polynesia, New Caledonia, Wallis and Futuna, French Southern and Antarctic Lands, Mayotte, Saint Barthelemy, Saint Martin, Saint Pierre and Miquelon (France), Aruba, the Netherlands Antilles (Netherlands), Faroe Islands, Greenland (Denmark), Guernsey, Jersey, the Isle of Man, Akrotiri and Dhekelia (Sovereign Base Areas), Bermuda, the Turks and Caicos Islands, Anguilla, the British Virgin Islands, the Cayman Islands, Montserrat, the Falkland Islands, the Pitcairn Islands, Saint Helena, British Indian Ocean Territory, South Georgia and the South Sandwich Islands (United Kingdom).*

This process included two of the countries, which are used as case studies in this book, and it will be seen that each country was granted different accession conditions, timetables and processes. Since the circumstances of each country differ, their application processes require specific actions and strategies to meet the EU's requirements. In fact, when this study was started in late 2002, neither Estonia nor Bulgaria where EU members. Since then Estonia achieved membership in May 2004 together with Cyprus, the Czech Republic, Hungary, Latvia, Lithuania, Malta, Poland, Slovakia and Slovenia. Thus, the EU's membership list jumped from 15 countries to 27 members. Its surface area increased by a quarter, and its population by one fifth, to 497 million.

2.3.1 General Accession Criteria

In 1999, at the Copenhagen European Council, the Member States took a decisive step towards the current enlargement, agreeing that *"the associated countries in central and eastern Europe that so desire, shall become members of the European Union"*.[37] Furthermore, the European Council provided a clear position for achieving accession i.e. *"Accession will take place as soon as an applicant is able to assume the obligations of membership by satisfying the economic and political conditions required"*.[38]

At the same time, the Member Countries designed the membership criteria, which are often referred to as the Copenhagen Criteria.[39] These are that membership requires a candidate country to achieve stability of institutions which will guarantee democracy, rule of law, human rights and respect for, and protection of minorities. The existence of a functioning market economy as well as the capacity to cope with competitive pressure and market forces within the Union were also prerequisites. Furthermore, they should have the capacity to fulfil the obligations of membership, which would include adherence to the political, economic and monetary aims of the union. Finally, it would also require the candidate country to prove that it has created the conditions for its integration through the adjustment of its administrative structures, so that European Community legislation transposed into national legislation is implemented effectively through appropriate administrative and judicial structures.

---

[37] See id.
[38] See id.
[39] The Treaty of Copenhagen was signed in 1993.

The impact of European Union accession on water utilities   17

All the above conditions have been summarized in the *acquis communautaire*.[40] In it the rules and regulations of the EU, the European Court of Justice and the Treaties are included. Adoption of the *acquis* alone, however, is not enough; competent institutions/bodies must be set up, or strengthened, to ensure that the *acquis* is properly implemented and enforced.

In the two case studies of Bulgaria and Estonia, both countries had implemented an aggressive programme to align national institutions to permit the full adoption of the *acquis*.

2.3.2 Drivers for implementing Accession Criteria

Countries aspiring to membership of the EU needed to have a clear mechanism to achieve accession. Thus, in December 1995, in Madrid, the European Council referred to the need to create *"the conditions for the gradual, harmonious integration of the candidates, particularly through the adjustment of their administrative structures"*.[41]

The Luxembourg European Council (December 1997) further emphasised that *"as a prerequisite for enlargement of the Union, the operation of the institutions must be strengthened and improved in keeping with the institutional provisions of the Amsterdam Treaty"*.[42]

In line with this basic philosophy of assisting the way in which Accession Countries are integrated into the Union, the European Commission developed a document called *Agenda 2000*.[43] In it, the Commission provides for various mechanisms to support Accession Countries. One of the areas of support outlined in this document was to assist Accession Countries to incorporate Community legislation into national legislation through appropriate administrative and judicial structures. This, in the words of the Commission, is *"an essential pre-condition for creating the mutual trust indispensable for future membership"*.[44]

---

[40] The *acquis* is also known as the legal and institutional framework by means of which the European Union implements its objectives.

[41] European Council, Madrid Proceedings December 1995, taken from http://europa.eu.int/comm/enlargement/docs/index.htm

[42] Fontaine, P, Europe in Ten points, European Commission (1998) N° cat. PD-13-98-257-EN-C at 17 Amsterdam Treaty.

[43] Agenda 2000 taken from http://europa.eu.int/rapid/start/cgi/ag2000/guesten.ksh?p_action.gettxt=gt&doc=DOC/99/16|0| RAPID&lg=EN

[44] Commission of the European Communities, *2001 Regular report on Estonia's Progress towards accession*, Brussels, 13.11.2001 SEC(2001)1744.

18   Public and Private Participation in the Water and Wastewater Sector

The Treaty of Nice[45] marked a new stage in the preparations for enlargement of the European Union and amended all existing Treaties. It aspired to include countries of Central and Eastern Europe, the Mediterranean and the Baltic.[46] It is important to note that the Treaty of Nice is only concerned with one crucial question: how can Europe function effectively when the number of Member States almost doubles?[47&48] Thus, what Nice did was to make a number of much needed improvements in the institutional framework of the European Commission. Some of these changes include,[49] amongst others, changes in

---

[45] Treaty of Nice amending the Treaty on European Union, the Treaties establishing the European Communities and Certain related Acts, 10 March 2001, Official Journal of the European Communities, 2001/C 80/01 [hereinafter, Treaty of Nice].

[46] The composition and operation of the European institutions and bodies were agreed in the 1950s when the Union only had six members (Belgium, Germany, France, Italy, Luxembourg and the Netherlands). The Union has since undergone five enlargements and now consists of 26 Member States. In addition, Turkey has been recognised as a candidate for membership.

[47] Protocol on Enlargement, Doc. SN 533/1, Brussels, July 10,2000, SEC (2001), taken from ICG website http://europa.eu.int/igc2000/sn533

[48] Memorandum to the Members of the Commission, Commission of the European Communities, Secretary General, Brussels, January 18 2001, David O'Sullivan,) http://europa.eu.int/igc2000/enlargement/s5-555.doc, at 2: Changes within the institutions during the enlargement process: First of all as it is not yet known exactly when and in what order the applicant countries will join the Union, the new distribution of seats in the European Parliament, the new composition of the Commission and the new definition of qualified majority within the Council are determined by the Treaty of Nice for a Union of 15 Member States. The Treaty restricts itself to setting out the principles and methods for changing this system as the Union grows.

[49] See id., 1–13, Some of the changes include:
Changes to the Composition of the European Parliament: the number of seats in the European Parliament will rise from 700 to 732 – see SN 533/1, p. 78;
Art 191 of the EC has been supplemented by a legal base which allows the adoption of European level political parties via the co decision procedure of the statute and particularly the rules concerning their funding;
The number of votes allocated to each Member State has been changed – see SN 533/1, p. 79 to reflect the more densely populated numbers of some Member States;
Changes in the composition of the Commission by imposing a ceiling on the number of members of in the Commission as well as increasing powers to its President (Art 217);
Changes in the Union's legal system including the distribution of responsibilities between the Court of Justice and the Court of First Instance (Articles 230, 232, 226 and 225);
Changes to the rules of the decision making process of the European Central Bank and the European Investment Bank;
Widening of the scope of decision-making process by qualified majority voting. There is a list of 27 provisions that have partially or completely changed or from a unanimity vote to qualified majority voting, e.g. judicial cooperation, measures to facilitate freedom of movement for citizens, industrial policy etc.
Under Art 7, if there are serious and persistent breaches of fundamental rights the council may suspend certain rights of the country concerned.

voting rights, instead of a unanimous vote, qualified majority votes, limits on membership to the Commission, and changes to relieve the burden on the European Court of Justice and Court of Auditors.[50]

In Santa Maria da Feira[51] and in Gothenburg[52] in June 2000 and June 2001 respectively, the Commission reiterated the importance of the capacity of Accession Countries to implement and enforce the *acquis,* and added that this required important efforts by the applicants in strengthening and reforming their administrative and judicial systems. To aid this, the Commission created a number of financial support instruments[53] to help Accession Countries fulfil the criteria for membership. At the same time the Commission must ensure that it monitors the Accession Country's ability to implement the *acquis* in its various aspects.

---

[50] The institutions and Bodies of the European Union: Who's who in the European Union? What difference will the Treaty of Nice make? Luxembourg: Office for the Official Publications of the European Communities, (2001) ISBN 92-894-0490-6.

[51] European Council, Santa Maria da Feira, Portugal Proceedings June 2001, taken from http://europa.eu.int/comm/enlargement/docs/index.htm

[52] European Council, Gothenburg Proceedings June 2002, taken from http://europa.eu.int/comm/enlargement/docs/index.htm

[53] *See* Presidency Conclusions Helsinki European Council, *supra* note 35. There are three pre-accession instruments financed by the European Community. These are: the *Phare* programme which provides institutional building and support for adoption and implementation of the *acquis* as well as support for economic and social cohesion; *SAPARD,* which provides aid for agricultural and rural support; and *ISPA,* which finances infrastructure projects in environment and transport. These programmes concentrate their support on the accession partnership priorities that help Accession Countries to fulfil the criteria for membership.

3
Public Private Partnership as an instrument to EU accession

3.1 PUBLIC PRIVATE PARTNERSHIPS: A POSSIBLE SOLUTION?

Faced with the challenge of meeting the Water Framework Directive, and in particular the growing investments required in the sector in order to comply with the EU's Water Framework Directive, governments of Accession Countries need to develop strategies to help achieve accession criteria.

Both Member and Accession Countries are increasingly interested in developing arrangements that will enable them to better meet the challenges

© 2009 IWA Publishing. *Public and Private Participation in the Water and Wastewater Sector: Developing Sustainable Legal Mechanisms*, by X Cledan Mandri-Perrott. ISBN: 9781789065299. Published by IWA Publishing, London, UK.

of developing basic infrastructure.[54] In particular, Accession Countries are characterized by a series of historical conditions that resulted from 50 years of a command economy. Public interest in cooperation with the private sector – a public private partnership – is driven by budgetary constraints as well as seeking increased efficiency, both in terms of service and allocation of finite financial resources.[55]

Public Private Partnerships can take many forms. In its simplest form, the term PPP refers to a structure of cooperation between a public authority or authorities and the private sector. Its primary aim is to fund, construct, renovate, manage and, or, maintain a given piece of infrastructure or the provision of a service.

In the context of the EU, PPP arrangements have a series of defining characteristics. PPPs are typically arrangements of relatively long duration, involving cooperation between the public partner and the private partner on different aspects of a planned project. Funding typically involves a combination of sources, both public and private which in some instances means complex arrangements between the various players. Notably, the role of the private sector is in the form of an 'economic operator', who participates at different stages in the project e.g. design, completion, implementation, funding. The public partner focuses primarily on defining the objectives to be attained in terms of public interest, quality of services provided and pricing policy, and it takes responsibility for monitoring compliance with these objectives.

However, if the private sector is to be involved effectively there is a need to balance the interests of the public and private sectors and allocate the risks between them fairly. This will permit the most suitable contractual form to be developed. Notably, PPPs are not a miracle solution, and so for each project there is the need to assess whether the suggested partnership can in fact add value compared to other more traditional contractual forms.

This chapter looks at the conditions which lead to considering the use of a PPP structure in the water and wastewater context, the various potential PPP forms as well as the conditions necessary to develop an effective PPP scheme.

---

[54] See, Commission of the European Communities, Green Paper on Public-Private Partnerships and Community Law on Public Contracts and Concessions, Brussels 30.4.2004, COM(2004) 327 final.

[55] The allocation of financial resources is not the only constraint that public utilities face in funding and or financing infrastructure projects. Infrastructure projects create a series of contingent liabilities (i.e. liabilities which are not immediate but for which adequate financial provisions should be made) which must be adequately assessed and accounted for.

3.2 CONSIDERATIONS FOR PPP IN THE PROVISION OF WATER AND WASTEWATER SERVICES

Public sector activities,[56] and in particular for water and wastewater infrastructure, are characterised by massive under-investment. A recent report from the Commission[57] estimates that for ten countries of Central and Eastern Europe alone, the cost of all investments needed to comply with the EU environmental requirements for drinking water supply,[58] wastewater and waste management exceed €120 billion.[59] At the same time private sector involvement is now seen as an option[60] to overcome the investment, technological and expertise gaps[61] facing many countries.

Given these characteristics and the inability of Governments to fund such massive investments, it has become apparent that new and innovative structures must be devised to 'bridge' these funding gaps. The most obvious option available to national governments is to turn to the private sector for help. As a result, between 1990 and 1997 the cumulative expenditure in new private sector capital for water projects in developing countries around the world increased from US$25 billion up to a staggering US$297 billion.[62] It is worth noting that private sector involvement is not simply a developing country option. Countries such as UK, France, Germany, Spain and Portugal, have seen significant increase in the participation of the private sector in the provision of water and wastewater services.

Moreover, it must be emphasised that a careful *balance between the interests of the public and private sectors* is needed. On the one hand, the public sector has a responsibility and obligation to provide services to its inhabitants and

---

[56] Communication from the Commission: *The Challenge of Environmental Financing in the Candidate Countries*, COM(2001) 304 Final, ELLIS Publications.

[57] World Water Council's, *Investing for the Water Future*, http://www.worldwatercouncil.org/vision/documents/chapter5.pdf

[58] *See* Water Framework Directive *supra* note 8.

[59] *See* Communication from the Commission, *supra* note 57 at 14. "But the Accession Countries are not alone in facing such bills, for example, the United States requires investments over US$325 billion, with the rest of the world's investment needs estimated at US$1.8 trillion."

[60] Labre, J., *The Role of the Private Sector*, Proceedings of an International Water Seminar, June 7–10, 1999, University of Dundee, Scotland.

[61] ADB: Developing Best Practices for Private Sector Investment in Infrastructure, at 4: The incentives associated with private, rather than public sector participation, may also lead to improvements in the operational efficiency of water supply utilities. http://www.adb.org/documents/books/developing_best_practices/water_supply/default.asp

[62] Johnstone, N., Wood, L., & Hearne, R., *Private Sector Participation in Urban Water and Sanitation*, (1999) 23 NRF 287.

ratepayers and on the other hand the private sector is entitled to receive adequate compensation for the risk of investing in, and managing, the utility as well as being recompensed for risking its private capital.[63]

Accordingly, the design of the structure for private sector involvement involves a clear definition of the roles to be played by the government and the private sector, the type of structure to be adopted for private sector participation, the applicable regulatory regime and its effects, as well as the specific current economic and political conditions.[64] Clearly, the type of structure chosen will leave an indelible mark on the provision of services and it is essential that an adequate legislative and regulatory framework be in place to ensure the sustainability of such a process. Regulation has to strike a balance between giving the private sector incentives to invest and operate efficiently, but at the same time protecting the interests and ultimate responsibilities of the public sector.[65&66]

The types of risks involved in the provision of water and wastewater services need to be adequately quantified, qualified and, most importantly, mitigated, as these are elements that will determine the success of a PPP project.[67] The research shows that all these risks must be properly understood. For example, according to Rivera,[68] risks to the private sector may be related to technical matters – due to lack of sufficient knowledge of the installations and the associated operations including necessary capital works,[69] commercial problems of not

---

[63] Bradbury, R, *Privatization of Natural Monopoly Public Enterprises. The Regulation Issue,* (1992), Policy Research Working Paper 862. World Bank, Public Sector Management and Private Sector Development, Washington DC.

[64] Cook, P., & Kirkpatrick, C., Privatisation, Enterprise Development and Economic Reform, (Cheltenham, Edward Elgar Publishing., 1998).

[65] Rees, J.A., Regulation and Private Participation in the Water and Sanitation Sector, (1998) 22 NRF 95.

[66] *See* Johnstone, *supra* note 63, at 34: The private sector is driven by profit, and if unregulated...can lead to under-investment in the human and social capital that is necessary for meeting basic infrastructure needs.

[67] Hemming, R., & Mansoor, A.M., *Privatization and Public Enterprises,* (1988), Internal Paper No 56, International Monetary Fund, Washington DC, at 2: Careful analysis of the risks involved is required at an early stage of the process, and the sharing of risks between the private and public sector should follow the guiding principle that whoever can control the risk best should assume it and should receive adequate compensation for doing so.

[68] Rivera, D., Private Sector Participation in the Water Supply and Wastewater sector: Lessons from Six Developing Countries, (The International Bank for Reconstruction and Development, The World Bank, Washington DC 1996).

[69] *Id.,* at 8:..."additional risk is that construction costs will escalate beyond what is planned because of unit price escalation beyond expectations and delays in construction timetable. As a rule, the construction risk is better borne by the private sector, because experience has shown that private companies are better qualified to execute investments within the budget and timetable envisaged".

being paid, or unable to recover costs, or tariff rates that do not allow full cost recovery with a reasonable level of profit, and financial risks which include currency devaluations and conversion of local and foreign currency, such was the case recently in Argentina.[70] Other risks include legal uncertainty related to the manner of resolving disputes, and political risks due to a change in government policy or the public sector's reluctance to increase tariffs. Conversely, risks to the *public sector* might be that services are not in accordance with agreed pre-defined standards and there may be political backlash.

What this demonstrates is that the allocation of risk between the parties should suit the relevant conditions in question. In the author's opinion it can be argued that in stable economies with reliable financial institutions and institutional frameworks, the private sector should be encouraged to take commercial risks and compete for credits in the financial markets. Conversely, it is unlikely that the private sector would be willing to bear full financial risk in countries with economies in development, rapid population growth, and low coverage of water and wastewater services, and where financial institutions and existing legislative and regulatory framework are still in their infancy.

3.3 PROJECT FINANCE, PPP OPTIONS AND IMPLICATIONS

Project Finance is a method of funding in which the lender looks primarily to the revenues generated by a single project both as the source of repayment and as security for the exposure. Typically PPP[71] projects are done on a project finance basis in which a private operator is granted a right and, in return for that right, has the obligation to provide a service. Under a project finance scheme, the right that is granted by the public sector is the right to expropriate, use, treat and sell water, in exchange for charging tariffs to consumers for such services. The types of PPP structures are defined later on in this chapter.

---

[70] *Id.*, at 8: The issue of currency conversion is an important risk, because revenues will be in local currency and part of the investments and borrowings will be in foreign currency. If repayment of borrowed money has to be done in foreign currency but income is in local currency, the private sector will be exposed to any change that may result between local income and foreign currency repayments.

[71] In the author's opinion there is no right or wrong PPP option. All structures described in this chapter are flexible and could potentially progress from less risky arrangements with no private investment, to more risky arrangements involving private investment as the two parties gain credibility and confidence in each other. Clearly, what is important to understand fully is that there is no exact form of PPP that fits all conditions exactly. The type of PPP structure chosen and designed must reflect the actual conditions and local circumstances.

Projects which involve Public Private Partnerships (PPPs) need to be 'bankable' projects. A 'bankable' project is one that combines a multitude of characteristics in a way which allows for the public and the private sector to share risks. In the EU context we look at PPP schemes in the water and wastewater sector. These projects need to fit within the regional/trans-national, national and local levels including compliance with EU framework and as appropriate EU legislation, local law etc. Additionally, financial viability and cost effectiveness are crucial factors of bankability as is funding appetite (either donor or debt funding). Typically projects which are considered as bankable include those projects which have characteristics that include clear contractual structures, a stable economic and financial basis and clear definition of roles, risks and responsibilities.

There is an array of options for involving the private sector, characterised by a variety of contractual forms. These contractual forms differ in the degree of responsibility and risk transferred from the public to the private sector.[72] Choosing the appropriate option depends on what the government in question requires from the private sector, i.e. the objectives to be satisfied by entering into such contractual agreements, and, as a result, how much control of the service it is prepared to relinquish. The following section examines the different forms of PPP based on the degree of private involvement in the development and ownership of the assets.[73] Throughout this section reference will be made to various examples where these forms of contract have been used. These examples draw upon the author's personal experience unless otherwise stated.

In the first group, ownership of the assets remains with the government or the public sector, whereas in the second group partial or full ownership is transferred to the private sector.[74] Each type of contract will be reviewed according to the percentage of private sector involvement.

---

[72] Lee, T.R., *Alternatives for Private Participation in the Provision of Water Services*, (1996) 20 NRF 333, at 14: The many options for private sector participation oscillate between one extreme of almost total public sector responsibility (e.g. leases) to totally private responsibility (divestiture) at the other.

[73] This classification is based on the four forms of private sector participation contracts generally accepted i.e. Service and Management Contracts, BOT and variants, Lease and Concessions. However, it must be noted that these forms are general categorisations and as such, not all contracts fall neatly into each category. Increasingly we are seeing variations in these types of contracts to suit the local needs. In Chapter

[74] Brook Cowen, P., Ensuring *Access to Water: Private Sector Participation and Services'*, Proceedings of an International Water Seminar, June 7–10, 1999, University of Dundee, Scotland. Choosing an Option: Governments have multiple objectives from private sector participation…technical and managerial expertise…improved efficiency…large-scale private investment…making the sector more responsive to consumers' needs and preferences.

The first group, where the assets remain in public ownership, include the following type of contracts:[75] Service and Management contracts, Lease arrangements and Concessions.

The second group, where the assets are partially in private ownership, include: BOOT (Build-Own-Operate-Transfer), variations to BOOT contracts such as BOT (Build-Operate-Transfer), BOO (Build-Own-Operate) and reverse BOOT, joint ownership or mixed companies, and outright sale, or partial divestiture.

These contractual forms will be enumerated below and the manner in which each contract allocates roles and responsibilities will be described.

3.3.1 Service Contracts

In a Service contract the public authority retains overall responsibility for the operation and maintenance of the system but contracts out specific, limited-scope services. It is important to note that in these types of contracts[76] the public authority bears all the commercial risk and must finance fixed assets as well as working capital. The private sector's responsibility is limited to managing its own personnel and services efficiently.

The types of activities that may be covered by Service contracts include: maintenance, emergency repairs, meter reading, billing and collection, upgrading of existing facilities or construction of new ones, and rental of equipment.[77] These contracts are for a short period of time, usually not exceeding two years, and are generally renewable.[78] The basic premise of these contracts is that they require little or no fixed investment on the part of the private operator. Efficiencies are effectively 'in built' due to the short contract period.

In the author's opinion, Service Contracts are an effective mechanism to bring about specific improvements. These have been used effectively by many public entities to improve, for example, billing and collections. In these schemes, such as with the Guatemala City Municipal Water Company, the utility entered a three year service contract with a private entity specialised in the issuing and collection of bills. Using a simple contractual structure, this was an effective mechanism to improve the financial standing of the entity.

---

[75] *See* Idelovitch & Ringskog, *supra* note 9. The same type of classification for PPP contracts has been used.

[76] Service Contracts are the simplest form of private sector participation.

[77] Examples of service contracts include the Mexico City contracts where the city was divided into four sections and contracts negotiated with four international operators.

[78] Bradburd, Ralph, *Privatisation of Natural Monopoly Public enterprises. The Regulation Issue.*'(1992) Policy Research Working Paper 862. World Bank, Public Sector Management and Private Sector Development, Washington, DC, Processed.

One of the main advantages of these contracts is that in large urban areas, for example, different firms can be contracted in separate geographical areas to deliver the same services. Multiple contracts ensure adequate competition and enable the water authority to compare costs and performance on an ongoing basis.[79]

However, it must be noted that if Government is to use this type of contract they will need to be prepared to manage the private sector, i.e. their role is to shift from that of execution to supervision.[80]

3.3.2 Management Contracts

In a Management contract the public sector transfers responsibility for the entire operation and maintenance of a system to a private company.[81] This gives the private company freedom to make day-to-day management decisions without assuming any commercial risks.[82] What is important to note in this type of contract is that the private sector acts in the capacity of 'agent' to the public sector, i.e. it acts at all times on behalf of the public authority. However, payment to the private sector may be contingent on collection of rates/tariffs from the customers.[83] Financial responsibility continues to be vested in the government or public authority and it has the responsibility to provide working capital and investment funds.

Payments to a management contractor are usually proportional to certain physical parameters,[84] such as improved efficiency, volume of water produced, improved collection rates, or reduction of unaccounted-for water.[85]

---

[79] *See* Idelovitch & Ringskog, *supra* note 9, at 15: A major benefit of service contracts is that payments to the contractor are linked to the work performed, instead of guaranteed wages paid to workers of a public utility. For example, a contract for reading meters would stipulate that the operator be paid a set amount for each meter read.

[80] *See id.*, at 17: For example, institutional reforms may be required to decentralise control, to provide technical assistance at a local level, to enforce quality and control standards, and to manage changes in staff.

[81] Delmon, J., Privatisation, London, Sweet & Maxwell, (2000), at 23: Management contracts are often used by multilateral agencies in order to inject special management skills into local water companies.

[83] *See* Bradburg, *supra* note 80, at 18: Under this arrangement, the contractor has no direct legal relationship with the consumer.

[83] The risk of collection of payment is passed on to the private operator.

[84] Yepes, G., and Dianderas, A., Performance Indicators. Water and Sanitation Utilities. 2nd Edition. Financial Indicators and Overview of Service Rates'. (1996) World Bank, Transport, Water, and Urban Development Department,, Washington, DC.

[85] *See id.*, at 56: Such a payment system creates an incentive for increasing productivity. In case the management contractor does not control the functions affecting productivity or quality, it is often compensated on the basis of a fixed fee.

The duration of these contracts is usually three to five years. Therefore management contracts do not entail large investments with long payback periods to be made by the private sector.[86] The intention is to act as a kind of 'interim' management prior to finding a more permanent and robust form that brings capital investment. Thus Management Contracts[87] may feature an option to convert to more intensive, longer-term PPP arrangements.[88]

Management contracts have not been as effective as originally hoped. From the author's personal experience, an example of a management contract that worked well, for the most part, was that of Trinidad and Tobago, where he was Director of a management contract that had a sound contractual structure and some very clear private operator performance indicators. The idea behind the management contract was to achieve real improvements through private management of the water company (the Water and Sewerage Authority of Trinidad and Tobago – WASA) enhanced by a targeted investment programme. However, at the time the management contract was designed and subsequently let, a loan from the World Bank was envisaged, specifically designated for the implementation of major capital investment involving major system rehabilitation. In reality, what happened was that the management contract was implemented for a period of three years but there was no capital programme as the World Bank loan did not materialise. The pipework and infrastructure required major rehabilitation of old and leaking pipework and structures, and this was not possible without the planned investment. Therefore, the ability of the management contractor to bring about significant operational and financial improvement was limited in aspects directly related to capital investment, such as continuity of water supply or reduction of leakage.

3.3.3 Lease Contracts

Lease Contracts have been commonly used in France and are also known as *affermage*.[89] In this type of arrangement a private operator rents or leases the

---

[86] Stiggers, D., Eyre, B., Graham, S., *The Trinidad Management Contract*, (1998), AWRA Conference Proceedings Paper.

[87] Examples of these include Amman, Jordan, Trinidad and Tobago and Guyana. The Guyana Management Contract has recently been awarded to Severn Trent Water International for a period of five years. It is too early to decide whether this contract has achieved the benefits that are expected under the contract.

[88] *See* Yepes & Dianderas, *supra* note 86, at 7: In many cases, management contracts precede leasing or concession contracts. In this case, their main purpose is to put the utility in order when the quality of service is poor or when accounting, consumer records, and information on the physical facilities are not reliable or available.

[89] *See* Idelovitch & Ringskog, *supra* note 9, at 15.

facilities from the public sector for a given period. In essence, the private sector is renting a permit for operation, maintenance, and management of the system[90] but the public sector is still responsible for capital expenditure for new projects, replacement of major works, debt service, tariffs and cost-recovery policies.[91] It should be noted that the public sector maintains the ownership of all assets and there is no transfer of ownership rights.

Lease contracts can be medium or long-term duration.[92] They usually last five to ten years[93] but can be extended for as long as twenty years. Lease contracts have a number of benefits as they allow for:

(a) Automatic periodic revisions of the rate being paid to the private sector through the use of price index formulas,[94]
(b) Periodic re-negotiations of the rates and cost index formulas on the basis of results achieved by the private sector (In this way, savings that the leaseholder has been able to achieve during the previous years can be passed on to consumers[95])
(c) Specific penalties that will apply in the case of poor performance.[96&97]

---

[90] The private sector, as the leaseholder, is responsible for all operation and maintenance functions, including offices, vehicles and spare parts, renewals and replacements, as well as for billing, collection, and financing working capital.

[91] Maksymiuk, Zbigniew, Transformation de L'Enterprise Communale en Societe Joint-Venture: Exemple de la Société Anonyme Saur Neptune Gdansk. (1994), Gdansk. World Bank, Washington DC. Processed. – at 32: In many cases, leaseholders pay the owners a rental fee sufficient to service the debt and finance part of the investment programme.

[92] Examples of these contracts include Guinea and Dar es Salaam Tanzania.

[93] See id., at 33.

[94] *See* Yepes & Dianderas, *supra* note 86, at 35: Payments to leaseholders are contingent with the difference between the tariff revenues collected and the operating costs. The contractor should be ensured a reasonable return on investment and receive incentives for maintaining efficient operations. The fact that the contractor depends on collections for revenue is an incentive to provide good service and establish good billing and collection practices.

[95] One caveat to this is that it relies on the ability and knowledge of the public sector to be able to identify and negotiate the potential for reducing costs.

[96] Lease contractors are usually required to put up a security deposit that can be called in by any public authority if performance is below the agreed standards.

[97] Triche, T., *Private Participation in the delivery of Guinea's Water Supply Services,* (1990), Policy Research Working Paper 477. World Bank, Transport, Water, and Urban Development Department, Washington DC: World Bank. – at 34: If, for example, a major goal for involving the private sector is to reduce pollution, penalties for not meeting quality standards can serve as an incentive backed by the contract. A common performance indicator is the level of unaccounted water.

Given the above factors, the risks involved in a lease arrangement tend to be limited as they generally allow the private sector to become more comfortable with the system and associated risks.[98] Accordingly, Lease Contracts can be used effectively for more extensive involvement in the future.

In the author's opinion, Lease Contracts run the risk of placing undue burden on the Government and its ability to raise funding. In the case of Dar es Salaam – Tanzania, where the author acted as senior advisor to the Government and the World Bank, the main issue was to ensure that the lease payments made were sufficient to cover some of the associated cost of the loan programme that was designed for capital improvements.

3.3.4 Concession Contracts

In a Concession,[99] the private contractor, or concessionaire, has overall responsibility for the services.[100] The distinctive feature of this type of contract is that the private sector is not only responsible for operation, maintenance and management, but it also has full responsibility to source funds for the capital investments needed for the expansion of services, either directly or indirectly (i.e. through banking or financial institutions). The fixed assets, however, remain the property of the government or public authority, but they are entrusted to the concessionaire for the duration of the concession contract and must be returned at the end of the concession period.[101]

The duration of Concession contracts is usually twenty to thirty years, depending on the level of investment and the payback period needed for the concessionaire to recover investment costs. If some capital expenditures have not been fully amortised by the end of the contract, there is usually a provision in the contract for the contractor to be compensated accordingly. The concessionaire

---

[98] *See id.,* at 15: When risks are limited, there is more competition from potential private partners, which benefits the public authority. In most cases, the public authority assumes the capital investment risk, and the leaseholder assumes the commercial risk. On the 13th of May 2005, the Government terminated its contract with City Water Services for the management of Dar es Salaam's water supply. The contract, in which the British firm Biwater was involved, was terminated on the grounds that the firm had failed to deliver services as per the contract. City Water claimed that the original bid documents were misleading and inaccurate. For detailed background see Tanzanian Affairs No. 82, Dr. Brian Mathew. September 1, 2005 at 12:51 am. Filed under Issue 82.

[99] Examples of these are the concession in Manila Philippines, Buenos Aires, Greater Buenos Aires and Mendoza in Argentina.

[100] Asian Development Bank Report, *Private Sector Participation*. http://www.adb.org/documents/books/private_sector_participation/default.asp

[101] *See* Bradburg, *supra* note 80, at 20.

retains exclusive rights for the duration of the contract. When the contract expires, all works and equipment are turned over to the public sector.

Under a Concession contract, the private sector is paid directly by the customer, based on the set contractual price.[102] Thus, the private sector assumes 'collection risk'. The private sector retains the balance of revenues after paying back any taxes and charges levied on customers by the public authority. If expenses exceed revenues, the private company suffers losses, which is the largest risk it assumes.[103] Penalties in Concession Contracts may be levied for failures to meet pre-defined targets.[104]

Concession contracts by their nature tend to be for long periods of time in excess of twenty years. Conditions change over such long periods and assurances made at the time of the signature of the contract might not necessarily hold true after a given period. An interesting example is what has happened with all the concessions in Argentina, where the Government defaulted when the foreign currency protection mechanism was exceeded due to the devaluation of the national currency (peso). In the author's opinion this is a case in point for the need to ensure that concession contracts are robust enough to allow the private sector to raise capital against these contractual structures, but that at the same time the contracts are sufficiently flexible to deal with unforeseen-changes in circumstances.

3.3.5 BOOT Contracts

Build Own Operate Transfer (BOOT) contracts[105] envisage the private sector financing, building, owning, and operating a specific new facility or system.

---

[102] Van den Berg, C., *Water Concessions*, http://www.worldbank.org/html/fpd/notes/– at 4: The advantage of combining responsibility for operations and investments in the same entity is that it provides an incentive to the operator to make efficient investment decisions, because their consequences will affect it directly. It also provides an incentive for technological innovations, because the operator will benefit directly from any improvements in efficiency.

[103] *See id.*, at 3: Water concessions create value by boosting service coverage and quality, and by improving the efficiency of utility operations…who wins, who loses from these concessions depends on how the contract is designed and regulated.

[104] *See id.*, at 5: Penalties in concession contracts are levied if the concessionaire fails to meet either the targets for service coverage or the quality of service specified in the contract. The penalty fines should be explicitly linked to each cause and should increase as the breach of contract becomes more serious. Penalties are generally paid to the regulatory agency.

[105] There are many of these structures in Chile and in Puerto Vallarta Mexico. Other examples of BOT schemes and their respective durations are in Mexico (5 years), Izmit – Turkey (3 years), Guatemala City – Guatemala (3 years).

The BOOT contracts have much in common with the Concession structure, but are generally used for a new facility or system, rather than operation and further development of an existing system

After a predetermined period of time,[106] ownership of the facility is transferred to the public sector.[107] The public sector is usually responsible for determining the demand for the service being contracted, and therefore for the size of the facility.[108] Demand is often guaranteed by the contracting agency representing the public sector in the form of a relevant authority, but problems may arise if there are differences between real and estimated demand.[109]

The duration of BOOT contracts is related to the time needed to cover the debt incurred, as well as the necessary time to provide adequate return to equity investors. At the end of the pre-defined period, the private sector transfers the facility back to the public sector.[110] As to the risk borne by each one of the parties, it is important to note that BOOTs may represent a substantial risk[111] for the private sector, particularly if there are no assurances that the investment output will be paid by the public sector, or that the quality of services that can be achieved will be uniform and according to design standards.[112]

---

[106] Martin, L., Private Sector Participation in Water Projects, P.Howsam (ed). (1996) Water Policy: Allocation and Management in Practice, London, E & FN Spon.

[107] *See* Idelovitch & Ringskog, *supra* note 9, at 19: BOOT contracts have not been used extensively in the water and sanitation sector but are becoming increasingly common in the power and transportation sectors. BOOT arrangements are attractive mostly for new plants that require large amounts of financing – for example, large water treatment plants or wastewater treatment plants – but they are not suitable for water distribution, or wastewater collection systems.

[108] Delmon, J., Water Projects: A Commercial and Contractual Guide, (2001), Kluwer Law International, Netherlands ISBN 90-411-1685-0.

[109] See id., 30.

[110] UN Bureau for Development Policy, *Public-Private Partnerships for the Urban Environment*, http://www.undp.org/ppp/gln/publications.default.htm, at 5: In some cases, provisions may be made for holding new negotiations to determine possible ongoing private interest in ownership or participation in operations.

[111] *See* Rivera, *supra* note 69, at 18: The lower risk of reverse BOOT, as compared to BOOT may encourage more private sector firms to participate. Reverse BOOT offers the advantages of efficient private sector operations and encourages the private operator to maintain the facility well because it expects to become the owner at some point in the future.

[112] *See* Idelovitch & Ringskog, *supra* note 9, at 17: Experience has shown that four issues require careful consideration:
First, the legal basis for private sector involvement has to be established in areas often reserved for the public sector.
The second issue pertains to the effect on financing caused by the size and length of time of these contracts. The first time a country undertakes the implementation of a BOOT transaction is usually longest. Subsequent projects may be expedited due to previous experience. Many BOOT projects

The traditional BOOT schemes may have a number of variations depending on the level of risk allocated to the parties. For example, one slight variation of the BOOT system is known as BOT. In this case ownership is transferred to the public sector as soon as the facility is completed, and the function of the private firm is only to construct and operate it.

One more variation is BOO, whereby ownership is not transferred to the public sector but remains with the private firm that constructs and operates the facility.

Another variation is used in countries where economic or political risks are high and where the private sector may be requesting very high risk premiums in return for their participation. In this case the public sector finances and builds the plant itself and then contracts a private firm to operate it over a long period of time. It is then possible for the private firm to acquire the plant gradually by paying the public sector an annual fee which usually covers the full debt service of the entire investment cost.

3.3.6 Joint Ownership

Joint ownership[113] is a good solution for situations where private sector participation is desirable but it is not possible to pass on all the risk of investment and operations to the private sector.[114] Under this scheme, a private sector firm and the public sector incorporate a firm under the country's existing

---

tend to be large and can cost millions of dollars. The large size and the long maturities required raises the complexity of the financing package and, often, the number of financiers needed to complete the package.

The third issue pertains to pricing and contractual arrangements. The risks are unique because of the normally regulated structure of the water industry. In some cases, pricing and contractual arrangements have to cover the project's foreign exchange risk as infrastructural BOOTs rarely have foreign exchange revenues. However, dividends to foreign equity investors, as well as debt service of external loans, need to be in foreign currency. Pricing is also difficult because it is often predetermined, and the private investor does not have the opportunity to recoup early losses by realising higher profits as industry conditions improve. These pricing arrangements are often based on pre-construction estimates of capital costs and project implementation times.

Fourthly, the level of tariffs and the quality of service provided can become particularly sensitive because, in most cases, the private company is providing a service directly to consumers.

[113] An example of joint ownership is Tallinn, Estonia.

[114] Munasinghe, M, Water Supply and Environmental Management: Developing World Applications, Boulder Colorado, Westview Press (1992), at 88: In some cases, it may not be feasible to pass full responsibility for investment and operations to the private sector, particularly during the initial stages of a sector reform programme. If the country environment is risky, capital investments may have to be separated from operations in order to attract private involvement in operations. In such cases, where it is desirable to maintain a higher degree of private sector participation than service, or lease contracts allow, joint ownership may be a good solution.

commercial code. The basic principle here is that, although the parties start with a fairly equal number of shares, the public sector has the option to sell part, or all, of its participation to the private sector in the future.

What becomes crucial in this type of agreement is the contractual agreement that regulates the relationship of the shareholders, i.e. the private and public sector. The roles and responsibilities are detailed in a shareholders agreement. Particularly relevant to this agreement is not only the mechanics for distribution of profits but the level of profits to be distributed between the partners.

Idelovich and Ringskog[115] argue that successful jointly owned companies can be expected to establish creditworthiness and to raise capital by floating bonds or issuing notes. This has the advantage of limiting public sector debt, an important sector objective. In the author's opinion, raising capital through bond flotation is unlikely due to the typical immaturity of the financial markets where these types of agreements tend to occur. In addition, supporters of joint ownership argue that the shares of the company would eventually be put on the local stock market and that pressure to operate efficiently will be exerted through stock market operations. The author considers this to be feasible in more advanced economies such as the UK where the public owns shares in the water and wastewater companies, but judging by the research carried out, there is yet to be an example in the developing world where this has been implemented effectively.

However, it must be noted that in countries with a weak regulatory tradition, the benefit of joint ownership is conducive to satisfying regulatory requirements since the public sector will be represented by a board of directors who will have a broad insight into the firm's operations.

3.3.7 Outright Sale/Divestiture

Some governments have opted for an outright sale of the water assets and wastewater services. The sale may be prompted by the public sector's desire to completely separate ownership from operations and maintenance.[116] It is also a way for the public sector to raise revenues. This has been a strategy used effectively in the sale of other utilities such as telecommunications and energy providers. However, it must be noted that in the water and wastewater sector its 'attractiveness' to the private sector will depend mainly on the rates or tariff they will be allowed to charge. It must be clearly understood that the assets themselves

---

[115] *See* Idelovitch & Ringskog, *supra* note 6, at 19.
[116] Vinter, G., Project Finance, (1998) London, Sweet & Maxwell.

have virtually no alternative value and that the business is modelled solely on its capacity to generate revenue and not on the value of the assets per se.

The most notable example of this is the sale of water companies in England and Wales. However, the particular political and commercial situation of the UK market at that time, with the general desire for full privatisation of public utilities, together with a well-established capital markets, may make this a difficult model to replicate elsewhere.

4
A summary on Economic Regulation and its implications on PPPs in the EU

4.1 BASIC PRINCIPLES OF REGULATION

As towns and cities developed, the provision of water and wastewater services passed from individual, to collective responsibility. Citizens began to organise themselves and form structures that were in charge of providing services to the community as a whole. Since these early beginnings, water and sanitation networks have grown in both size and scope. Associated costs and resource requirements have increased according to the complexities involved. Modern water and wastewater systems require large and continuous capital investment. The need for realising a return on this investment would intuitively seem to conflict with policy goals including universal availably and affordability. Left to

© 2009 IWA Publishing. *Public and Private Participation in the Water and Wastewater Sector: Developing Sustainable Legal Mechanisms*, by X Cledan Mandri-Perrott. ISBN: 9781789065299. Published by IWA Publishing, London, UK.

its own devices, a naturally competitive marketplace for water and wastewater services could result in redundant infrastructure in some areas and no infrastructure in others. Profit incentives for private investors and customers' ability to pay would probably not result in a socially optimal water and sanitation network. Because these services are vital to quality of life and economic development, the case for public intervention is quite clear.

In most countries, specific regions are served with water and wastewater services provided by a single entity. Structuring utilities as 'natural monopolies' helps to achieve universal accessibility and affordability and concentrate investment into one single infrastructure network. Inevitably, citizens have no choice but to receive service from this sole provider. The resulting combination of vital services and monopoly pricing power gives obvious cause for concern. Public ownership and control would represent the simplest method for protecting the interests of citizens. However, as discussed in previous sections, public ownership is not without limitations and failings.

Many cities require financial resources and implementing capacity that public sources simply cannot supply. This, of course, is one reason that private sector participation can add value and result in net social benefits. Unsupervised private monopolies would nevertheless be undesirable. Economic theory predicts that monopolists could set prices above a socially desirable level where the quantity of service supplied would also be sub-optimal. In contrast, citizens and policy makers would prefer pricing equivalent to the 'marginal cost' of supplying services.

Marginal cost is the additional cost associated with producing one more unit of output and is an important concept tied together with overall economic efficiency. Wisely allocating scarce resources implies that customers should pay the true costs for what they consume. Pricing below marginal cost leads to overconsumption because the benefits of each additional unit of service outweigh its expense. Pricing above marginal cost leads to under consumption because the expense of each additional unit outweighs its benefits. Competitive market forces naturally compel firms to price near marginal cost where the corresponding level of public consumption is also optimal. Unfettered monopolies prevent this phenomenon from occurring.

Providing capital intensive public services presents an interesting challenge. On one hand, monopolies represent the best way to achieve universal access at the lowest investment cost. On the other hand, monopolies left unchecked would not set price or supply services in an efficient manner. Reconciling these conflicting interests requires regulation to allow for efficient investment while also compensating for the absence of market forces.

The preceding discussion should not imply that regulation exists only to ensure economic efficiency. Other goals of sound regulation include:

1. Financial viability: Private firms expect reasonable returns on capital invested. Intelligent companies take risks only in accordance with the potential for rewards that may follow. When anticipated rewards are insufficiently small for some amount of risk, firms will not commit scarce resources to a project. Continuous investment is critical for maintaining quality water and wastewater services. Return on capital is also an important incentive when justifying system expansion to meet the goal of universal access. Effective regulation therefore should attempt to preserve investment incentives at some appropriate level. This threshold should approximate the 'opportunity cost' of scarce funds. For certain levels of risk, investors must receive a given level of return in accordance with fair market rates. Allowing firms to recover normal returns from their investments is a critical regulatory goal because it provides the incentives and corresponding investment required for maintaining good service quality and expanding the water and wastewater network to all citizens.

2. Operational efficiency: As previously discussed, water and wastewater services are socially beneficial and should ideally be priced at their respective marginal costs to optimize both consumption and supply. Reductions in the marginal cost of services allows for additional, yet still optimal, consumption. There is, therefore, clear benefit to reducing costs and increasing the efficiency with which a regulated monopoly supplies these services. Creating incentives for private sector efficiency requires some potential for reward in order to incentivize progress in making improvements and further cost reductions. Sound regulation should allocate some portion of the benefits from reduced service costs back to private partners as an incentive to improve operations and make them more efficient.

3. Dynamic efficiency: Rivals in competitive markets will undertake investment and formulate plans to accommodate future demand in order to retain or increase market share. In the absence of competition, regulation should stimulate private monopolies to still plan and invest for future growth as unsupplied demand would again be undesirable.

4. Distributive justice: Access to clean water and adequate sanitation is a fundamental human right. Accordingly, policy makers should seek that those with access to services can also afford them. Achieving this goal may require some form of subsidy in many cases. These subsidies often come from customers with greater ability to pay for services (either directly through a 'cross subsidy' arrangement or indirectly through tax monies). In this way the structure of prices is consistent with paying capacity.[117] Balancing the inherent conflict

---

[117] Estache et al. 2003.

between distributive justice and economic efficiency is another goal of good regulation.[118]

4.2 REGULATION OF PUBLIC AND PRIVATE SERVICE PROVIDERS

As discussed above the provision of water and sanitation is an essential service, and has strong natural monopoly characteristics. This makes it important to control service providers and make them accountable in some way, regardless of whether the provider is publicly or privately owned and operated. Private providers are generally regulated by autonomous agencies or contracts, or both. Regulation is considered necessary because, without some form of constraints, private providers may act in a way that is not in the public interest or make unreasonable profits by overcharging and under delivering.

Economic regulation can be defined as the legal controls placed on service providers in order to overcome the problems inherent in an essential, monopoly service. Economic regulation sets the formal rules and associated organizations that set, change, monitor, and enforce allowed tariffs and allowed service standards for water providers.[119] Ehrhardt, et al, further defines economic regulation as mimicking the pressures that competition provides in other markets. In other words it can help to stop tariffs from increasing above the level required to recover reasonable costs. It will also help to set and enforce service standards and to ensure that those standards are achievable against set tariffs.

It is important to point out the need to balance the level of tariff and the associated level of service, or the standard, that one seeks to achieve. There is no point in setting service standards which require massive investments that cannot reasonably be recovered by the service provider. If it is a private provider, tariffs will need to cover its reasonable costs i.e. the cost of operation and maintenance required to provide the service, as well as the cost associated with borrowing capital for making the necessary investments, and providing a reasonable return (or profit) to the shareholders of the service provider. Alternatively if the service provider is a public entity, recovery of reasonable costs should in, principle, seek to set tariffs which would cover these costs. However the key difference is that the cost of capital for governments is typically significantly lower than that

---

[118] Hantke Domas, Michael P., Economic Regulation of Public Utilities with Natural Monopoly Features, September 2005, University of East Anglia.

[119] Ehrhardt, Groom, Halpern, and O'Connor, Economic Regulation of Urban Water and Sanitation Services: Some Practical Lessons, Water Sector Board Discussion Paper Series, Paper No. 9, April 2007, World Bank.

obtained by the private sector, and those public borrowing benefits can be passed on to the customer.

Public provision should not per se imply that the sector is being subsidized in an un-transparent manner. Even if there is a public provider, the surplus after operations and maintenance, as well as government's cost of capital, should be retained by the service provider and reinvested. After all, one of the key characteristics of the water and wastewater sector is the fact that it is highly capital intensive. For this reason economic regulation should seek to set rules to ensure that a service provider is able to continuously invest to achieve sustainability of service. It is important that tariffs are set at the adequate level: not too high so as to allow excessive profits or surpluses, and not too low so as not to allow adequate cost recovery. Notably, it is important that government does not bear the fiscal burden of providing either implicit of explicit subsidies to the sector.

4.3 REGULATION, NATIONAL POLICY AND THE EU LEGISLATIVE CONTEXT

As discussed above, economic regulation focuses primarily on setting, monitoring, and enforcing rules on tariffs and service quality. Yet depending on the legal structure of a given country, its legal tradition and other factors, economic regulation may overlap with other areas of regulation and policy. In this book we are only concerned with the main principles of regulation and their implications with respect to EU policy and particularly how economic regulation can be 'woven into' a PPP regime. In this context, it is then up to each Member Country to apply its own rules as to the specific manner in which it seeks to undertake its regulatory functions. It is up to each Member Country to decide whether a particular regulatory function should be considered part of economic regulation or dealt with in another way. For example, setting drinking water standards may be considered part of economic regulation, or it may be dealt with as an aspect of health or food safety regulation. Service coverage may be governed by regulatory requirements on the provider to connect more customers, or by subsidies for service expansion, or both.

Another form of exercising regulatory influence may be through the use of powers granted at law to instruct providers to do certain things, and to enforce these instructions through penalties or providing other forms of incentives. All in all, what is important under an EU context, and in particular for EU Accession countries, is that whatever form of regulation is chosen it is designed in such a way that it is consistent with EU law. After all, the EU is explicit in allowing each Member Country the discretion to implement its own sector laws so long as they are coherent with EU law and specifically in the water and wastewater sector with the EU Framework Directive, EU Competition Law, principles for PPP and the EU Utilities Directive, amongst others. Of particular relevance are the implications of State Aid, as EU Competition Policy recognizes the need in some instances for State Aid. A Government may choose to influence providers through subsidies such as output-based payments for extending service to poor households, or by providing specific tax incentives. All this needs to be set out in a comprehensive national set of rules together with the presence of adequate institutions to deal with the enforcement of these rules. This is all discussed in the next chapters.

In all this it is important to recognize that there are problems that economic regulation cannot solve, or at least cannot solve alone. A good example of such limitations is the need for a country to develop an integrated approach to ensure services remain affordable to low-income households, or that there are sufficient incentives placed on the service provider to increase efficiency, capital expenditure planning, and achieve low payments by government departments and agencies. Thus, even when governments choose to expand the role of regulation beyond simply setting, monitoring, and enforcing rules on tariffs and service quality, supporting regulation with other policy initiatives is generally necessary.

The EU context allows for each country to develop its own approach to the water and wastewater sector. Accordingly, each Member Country has to develop a set of national rules that overall provide for the correct implementation of the EU Water Framework. However, setting these 'rules of the game' should combine not only regulation but also a national framework that works hand in hand with economic regulation. For example, the overall policy making function should be developed to ensure that the Member Country sector objectives and principles, as well as the actions to achieve these, that are clearly set and consistent with EU policy. Uniformity with EU Utilities directive, as well as the principles of good economic policy, should seek to define the extent to which consumers or taxpayers should pay for water services. Furthermore, specific national legislation may prescribe ownership rights related to water, assets etc. including who is responsible for service provision, and how the owner exercises control over the utility's management. Member Country governments need to ensure that these factors and the regulatory design are aligned with each other. Furthermore, the regulatory regime must be coordinated with other bodies involved in water sector activities. It is important that each Member Country brings together the different bodies that deal in a direct or indirect way with the water sector. It must ensure adequate coordination of public expenditure priorities. Consistent and appropriate policy decisions and implementation plans are key to achieving a balanced approach to the development of the water sector.

Governments embarking on water and sanitation sector reforms need to decide how providers will be given better incentives to deliver the services that people want, at a reasonable price. This is where regulatory design becomes important. The regulatory system should include a combination of rules and organizations (including appropriate Member Country institutions) that are able to help deal with the country's key regulatory challenges, the inter-linking nature of Member Country law with EU law, and how these factors affect the manner in which national institutions can work. Brown et al[120] suggest that the starting point for any national regulatory framework should start with a thorough understating of the underlying problems and objectives in the sector. A deep assessment of whether existing regulatory systems are well suited to achieving the Member Country objectives, as well as identification of the specific regulatory functions needed to achieve those objectives, will be key. This assessment will help a Member Country to decide, in light of existing institutions, which legal instruments are best suited to embody the regulatory rules, and which organizations are best suited to perform the regulatory functions.

4.4 DEVELOPING A REGULATORY FRAMEWORK FOR EFFECTIVE PPP

The use of water resources and the manner in which wastewater is treated and disposed of into the environment needs to be well managed. When introducing the private sector into the provision of water and wastewater services, it is important to define clearly the roles and responsibilities of each of the parties involved. The suggestion is that the private sector should be responsible for providing the service, i.e. the operation and management (including sometimes financing of the sector's needs), whilst the public sector's role shifts to one of ensuring protection of this vital resource and regulating the manner in which the services are being provided.[121]

It is also important to recognise the potential for harm from bad regulation. In particular, regulation that oversteps it bounds can severely hinder investment, limit development in the sector and could ultimately amount to expropriation. This may occur when regulatory changes deprive private firms of the potential benefits that would otherwise accrue from an asset or other investment. It is

---

[120] Ashley C. Brown, Jon Stern, and Bernard Tenenbaum with Defne Gencer, Handbook for Evaluating Infrastructure Regulatory Systems, 2006 The International Bank for Reconstruction and Development/The World Bank.

[121] *See* Brook Cowen, P., *Competition and Regulation in Water and Sanitation,* Proceedings of an International Water Seminar, June 8–11, 1998, University of Dundee, Scotland.

essential that the framework under which the provision of such services are to be rendered is clearly defined beforehand within the context of existing property rights legislation, so that risks involved may be allocated to the appropriate party. This regulatory framework for including the private sector in the provision of water and wastewater services effectively sets 'the rules of the game' and its form may vary depending on the type of contract structure.

Typically, when the private sector is involved in the provision of water and wastewater services, there is some form of contractual relationship that enshrines the roles and responsibilities of each party. Notwithstanding the contractual form chosen, the framework needs to create a business approach clearly focussed on meeting the requirements of its customers, i.e. quality of product and customer satisfaction. Similarly, regulation must not only protect the customers from monopoly abuses but it must ensure some level of risk protection for the private sector, particularly in relation to investment.[122] In PPP contracts such as BOTs, Concessions or Divestitures the private sector has a long-term investment commitment that requires assurances that the project's structure will allow adequate recompense for repayment of the loan principal and interest.[123] Private sector participation needs an enabling set of rules not only of a regulatory nature but also from a political, legal and economic perspective.[124]

In essence the regulatory framework devised must have the following key design characteristics[125] that provide a practical set of rules, derived from the political, legal, economic and public perspective.

---

[122] World Bank Tool-kits, *Options for Regulating Water and Sanitation*, http://www.worldbank.org/fdp/wstoolkits/kit1/annex4.htm at 9: Regulations must not be designed just to protect public interests, but must also create the conditions under which private firms can operate effectively and efficiently.

[123] Sacerdoti, G., *Private Infrastructure for Development: Confronting Regulatory Risks*, International Conference Proceedings, Sept. 8–10, 1999, Rome, Italy, http://www.worldbank.org/riskconference/papers.htm, at 6: Foreign direct investment in infrastructure projects in developing countries are exposed to higher risks than other projects, due to their dependence on weak governmental regulation and the long term commitments that are inherent in such investment. Investment in infrastructure expose investors to major risks, since the projects involved tend to be large and their costs can only be recuperated over long periods of time.

[124] Ganiage, D., *Governance Dimensions of Regulatory Frameworks for Private Sector Development*, http://www.adb.org/Documents/Conference/Seminar_Governance_Frameworks.htm.

[125] The basic characteristics for a regulatory framework which the author believes are essential for the proper and sustainable use of a PPP structure in the provision of water and wastewater sector have been compiled after extensive discussions with members of the PA Consulting Group's Private Sector Advisory Group, in particular Mssrs David J Stiggers and Dr Kenneth Rubin, over a period commencing from May 2000 to January 2002. Consultation has also been made with the Director of the Public Sector and Infrastructure Sector, Latin America Mr Abel Mejia during a series of interviews in June 2001 and subsequently during December 2002. In addition, Mr Christopher

4.4.1 Predictability

Where there is a well-defined regulatory framework, the 'rules of the game' are clear to all parties and should be established in the contract itself enshrined in applicable legislation. Successful PPP structures rely heavily on adequate allocation of roles, rights, responsibilities and obligations between the parties within the contractual structure. However, Accession Countries do not have the luxury of well-established regulatory bodies and supporting legislation, therefore the contract needs to specify and allocate the risks involved in both the public and private sectors, besides the commercial contract arrangement between them.[126] This is especially important when property rights are poorly defined or inconsistently upheld. In effect, the contractual agreement between public and private parties serves as substitute for legal consistency.

4.4.2 Adequate alignment of the PPP contract with national law

Normally, specific laws should establish the broad principles of a regulatory policy and set national service standards. In developed countries such as the UK, this basic alignment between the PPP contract and national law is the norm.[127] Therefore the PPP contract complements such national law, whilst at the same time covering the more specific commercial details for the specific locality. This means that the PPP contract is predictable and enforceable, i.e. certain basic safeguards particularly relating to the protection of public interest should be incorporated in the contract. Such mechanisms will permit a predictable

---

Jennings, Senior Water Specialist of the Inter-American Development Bank was interviewed in July 2001. Numerous communications, via telephone and e-mail, have been made from May 2000 to February 2003. As a result of these investigations and discussions with leading regulatory experts and based on the literature review in this chapter, the following key issues of predictability, alignment with national law etc. provide a summary of what the author believes are the most salient characteristics required for the design of a regulatory framework. His own international experience which includes the setting up of the national regulator for Estonia, the regulatory working paper presented to the Prime Minster of Montenegro in December 2001 and the development of the regulatory commission in Sofia, Bulgaria has also been brought to bear in proposing the key design characteristics.

[126] *See* World Bank Toolkits, *supra* note 124, at 14: ...in order to mitigate problems encountered later on in the project's life...often a private sector contract contains many, or most, of the provisions governing regulation of the private sector in the contract.

[127] Roy, D., *How can Regulation be made Effective*, Proceedings of an ABD Conference, Geneva, Switzerland, April 28, 1998, at 3: In the USA and the UK regulation is well understood, has established rules of business, and has taken deep root.

regulatory framework and adequate allocation of risk to "the party that is best placed to mitigate that risk".[128] As the risk factor will be fully understood, this will give the private sector confidence to invest.[129] Developing countries are beginning to appreciate the need for such enforceable legislation to ensure a balance between the private and the public sector risks and benefits.[130] Accession Countries have also learnt this key lesson and have drafted enabling legislation as part of their accession criteria.

However, it should be noted that one of the key worries for Accession Countries[131] is that there is insufficient provision in national law to deal effectively with the items mentioned above in the private sector proposals, particularly in concessions.[132] This is a topic very much under current European Union consideration to strengthen the law regarding private sector involvement in general. It is felt that under current national law Governments are exposed to the possibility of private sector manipulation and potential monopolistic abuse. Not only are the national laws too weak but they also vary in their consideration of private sector involvement, creating a somewhat worrying level of legal ambiguity across the EU as regards such contracts.[133] The legal treatment of contracts currently varies significantly from one Member State to another. Such

---

[128] *See* World Bank Tool-kits, *supra* note 124.

[129] *See* Sacerdoti, *supra* note 125, at 13: If the assurances are weak...risks will be perceived to be high, then the water tariffs will also have to be higher...governments that do not put predictable and sound regulatory structures in place will have great difficulty achieving private sector participation in water supply.

[130] Wragge, S., *Public-Private Partnerships*, http://www.atse.org.au/publications/focus/focus-wragge.htm, at 2: South Africa has embarked on a program of developing a 'PPP-friendly' program of legislative and regulatory development to ensure that a common and efficient approach is taken to the use of PPPs.

[131] Barlow, M., *Private Drain on Public Water*, http://www.theglobeandmail.com/hubs/national.html, at 1: ...governments have very few laws or regulations to protect their water systems. Most haven't even begun to address the issues of privatisation...yet while they leave their water resources unprotected, they are actively negotiating and signing investment agreements...

[132] Economic and Social Committee, Strengthening of the Law Governing Concessions and Public/Private Partnership Contracts, 2001 O.J. (C 14) 91.

[133] *See* Introduction to the new EU Water Framework Directive, *supra* note 176, at 15: There have been many expressions of support for a European Directive on the subject aimed at ending unnecessary legal ambiguities and creating a harmonised legal basis for these contracts. National laws are not at present harmonised, and the rights and obligations they provide and their treatment of PPPs varies widely. It is not acceptable that the award and execution of this kind of contract should vary to such an extent within the Member States...The concepts of PPP must be clearly spelt out in European Law so that private investment in public facilities are not discouraged.

ambiguity across the board creates uncertainties for the private sector when dealing with different States, which is hardly conducive to predictable regulatory frameworks.

4.4.3 A well defined PPP contract

A well-drafted contract, specifically underlining the parties' duties and obligations, with recourse to strong and enforceable legislative safeguards, makes for an effective regulatory framework. Such predictable regulation will help ensure that private sector involvement in water is effective for both the private and public sectors.[134] Ensuring predictability also requires limitations on regulatory discretion where appropriate. Nascent agencies (as exist in many accession countries) may be particularly prone to whimsical changes in policy. Clearly defined contractual mechanisms that reduce regulatory discretion are essential to providing the security needed for reasonably priced private investment. As institutional capacity evolves along with transparent mechanisms for oversight, future PPP agreements can allow room for greater discretion.[135]

4.4.4 A degree of flexibility to deal with long contract periods

Most of the PPP contracts involve the parties over long periods of time sometimes in excess of 20 to 30 years. Due to this lengthy period, when drafting such PPP contracts or subsequent regulatory frameworks, it is not possible to make provision for all future risks.[136] Therefore it is important to provide a degree of flexibility to allow for the parties to the contract to react to changes of circumstance as the contract evolves. Moreover, to ensure no party tries to take any form of unfair advantage when changes to the contract are required, the regulatory body must be impartial in order to ensure that the original duties and obligations are maintained.[137] The function of the regulator is dealt with below.

---

[134] *See* Sacerdoti, *supra* note 125.

[135] Eberhart, Groom, Halpern and O'Connor, *Economic Regulation of Urban Water and Sanitation Services: Some Practical Lessons.* The World Bank, April 2007.

[136] *See* Roy, *supra* note 129, at 97: Social, economic, technical, political, demographic and legal conditions will likely change during the period of the project.

[137] *See* World Bank Toolkits, *supra* note 124, 'Is the Regulatory Framework Conducive to Private Sector Participation'.

4.4.5 Independence, accountability and transparency of regulation

Within the necessary independent and fully competent regulatory body, the regulator must not only ensure that the contractual and legal obligations are adhered to but also that the interests of all parties are upheld, at all times.

To do this, three main criteria need to be satisfied: *Independence, Accountability and Transparency*. The regulator must be independent in the sense that he is not influenced in anyway by any of the parties in the project;[138] he must be subject to supervision and accountable to some other responsible, external body; he should be obliged to explain and justify processes, actions taken, procedures and decisions made, to all concerned. Most importantly, the regulator must be seen to be totally impartial and purely working towards ensuring that the interests of all parties are met.[139]

Lastly, it is important that any regulatory framework intelligently accommodates the political situation at hand. Attempting to shut politics out of the process entirely may often result in exactly the opposite outcome when public pressures inspire rash actions. Savvy planning will provide a mechanism for conspicuous public involvement while preventing any kind of 'regulatory taking' or outright expropriation. For example, where tariff levels are politically charged issues, it may be wise to engineer separation between tariff regulation and operator revenue. Incorporating performance based subsidies or other mechanisms designed to manage operator behaviour can preserve incentives for efficiency and cost reduction while also maintaining the integrity of the regulatory regime.[140]

Annex 2 provides the full framework for analysis.

---

[138] *See* Bradburg, *supra* note 80, at 100: Regulators need to be isolated from political interests and other interest groups...a regulator should be appointed, based on prescribed and transparent technical and commercial criteria and not political considerations.

[139] Bitran, E., and Serra, P., *Regulation of Privatised Utilities: Lessons from the Chilean Experience.* (1995) OECD Advisory Group on Privatisation, Eight Plenary Session, Paris, October 30 and 31. Paris. OECD, at 3: The regulatory arrangements created must be capable of credibly striking a fair balance between investor and consumer interests. A basic requirement for this to be effective, the regulators should have a high degree of independence from both political interference and from the companies that they are regulating.

[140] *See* Ehrhardt et al, *supra* note 137.

5
European Legal framework and its implications on PPP

5.1 EU'S APPROACH TO PPP

The need for financing infrastructure services and in particular the water and wastewater sector is a challenge being faced by a number of governments seeking accession to the European Community. Moreover, Public authorities at all levels, including Member Countries, are increasingly interested in developing specific arrangements for cooperation with the private sector. Public Private Partnerships are one of the methods available, and as discussed under the previous chapter, there are various contractual forms available to undertake PPP. The choice of one of these forms is dependent on what a public authority seeks to achieve. Notably, within Community law there are no prescriptions at law for such relationships nor, in fact, whether a public authority should enter into a PPP. The decision to provide an economic activity is squarely in the hands of

© 2009 IWA Publishing. *Public and Private Participation in the Water and Wastewater Sector: Developing Sustainable Legal Mechanisms*, by X Cledan Mandri-Perrott. ISBN: 9781789065299. Published by IWA Publishing, London, UK.

the public authority. If it wishes to delegate a part or the whole of this activity to a third party it is the prerogative of the public authority.

However, any delegation of responsibility may be the subject of Community law on procurement. Accordingly, the form of delegation through a PPP instrument needs to be examined against the basic principles of EU law and the framework it establishes.

In this chapter we review the impact Community law may have on developing PPP structures. In the absence of a uniform legal framework for PPPs at European level and in view of the heterogeneity of national laws the question arises, whether uniform provisions are needed to safeguard transparency and effective competition between market participants. The EU has decided against the creation of a separate legal regime for PPPs (although it has issued a number of papers providing basic guidance for PPP implementation). The EU also considers that, as a matter of principle, the law on public contracts needs to be applied whenever a private partner is to be selected.

The main purpose of Community law on public procurement is to create an 'internal market' in which free movement of goods and services can take place. The intent is to ensure the right of establishment and the right to provide services (Article 43 and Article 49 of the EC Treaty)[141] are respected. The implication of this is that, under Article 48, the principles of equal treatment, transparency and mutual recognition must be assured. Value for money must also be obtained when a public authority buys products or mandates third parties with performing services and/or works.

Given the above issues, this chapter starts with a general examination of the general principles of EU law in terms of competition and procurement. It follows that the impact of the EU Utilities Directive, the implications of public involvement and ownership, and state aid must be examined. General principles of EU Competition law must also be understood and a simple test structure is presented to exemplify the legal concepts behind EU thinking in this regard.

Having understood the basic principles of Community law, it follows that the specific rules applicable to PPP in the EU context must be understood. The selection of a private partner under a given PPP structure derives firstly from the definition of the contractual relationship which the party enters into

---

[141] Under Community law, any act – whether it be contractual or unilateral, whereby a public entity or authority entrusts the provision of an economic activity to a third party is subject to the principles of Community law related to the creation of an internal market. In this respect, the rules and principles governing public contracts and concessions apply to any activity i.e. any activity which consists of providing services, goods, or carrying works in an internal market, even if tehse services, goods or works are intended to provide a "public service" as would be the case of a third party providing water and wastewater services under a PPP arrangement.

European Legal framework and its implications on PPP 51

with a public contracting authority (see previous chapter for contract types). Under Community secondary legislation, any contract for pecuniary[142] interest concluded between a contracting body and a private partner in the form of an operator, which has as its objective the execution of works, or the execution of works and services, is designated as a "public works or public services contract". The concept of "concession" is defined as a contract of the same type as a public contract except that as consideration for undertaking these works, a right to exploit has been granted or a combination of a right and a payment is made. Thus, under this section of the chapter we detail the manner in which PPP structures can be developed.

Given that our objective is to develop PPPs in the water and wastewater sector, it is contingent upon us to ensure that the PPP structure devised is consistent with relevant sector practice and legislation. The EU's framework for the sector is enshrined in the Water Framework Directive (WFD). The WFD is an inclusive approach to managing water as it flows through catchments from lakes, rivers and groundwaters, to estuaries and the sea. The WFD creates a framework for all future actions by the EU on water policies.

So, in this chapter we also examine the Water Framework Directive, its evolution as well as its main principles and objectives. We delve into the key principles behind the application of the WFD and its *managerial* and *environmental* objectives. Under each we identify the most salient issues as they apply to Accession Countries. These key principles, together with the principles of competition as they apply to PPP, will be utilised later on as the basis for the development of an analytical framework for examining more rigorously the provisions of the WFD against a real life situation in two Accession Countries: Bulgaria and Estonia.

5.2 RELEVANCE OF EU LAW IN THE WATER SECTOR

Throughout this book the possibility of using the private sector in a very definite and positive role to support the EU accession process has been emphasised. Therefore we all so look at the relevant provisions of Community law that may apply to the water and wastewater sector will be highlighted. In so doing, the implications on procurement and competition need to be examined. Community law on public procurement seeks to create an 'internal market' in which free movement of goods and services can take place. The principles of equal

---

[142] This refers to contracts that involve payment or some form of financial consideration for the provision of services under the contract.

treatment, transparency and mutual recognition must be assured. Value for money must also be assured. However, given the need to simplify and modernize procurement procedures of entities operating under the water, energy, transport and telecommunications sectors, a 'Utilities Directive'[143] coordinating the procurement procedures of entities operating in these sectors has been developed. This Directive is specific to the sector and a distinction has thus been made between these sectors and other forms of public procurement – the 'Public Sector Directive'[144].

5.2.1 EU Utilities Directive

The EU Commission recognised the need to simplify and recast the procurement Council Directive 93/38/EEC of 14 June 1993 coordinating procurement procedures. The new Utilities directive of 31 March 2004, is based on Court of Justice case-law, in particular case-law on award criteria. One major reason for the introduction of rules coordinating procedures for the award of contracts in these sectors is the variety of ways in which national authorities can influence the behaviour of these entities. This influence may be exerted through a participation in their capital and/or through representation in the entities' managerial structure. Another main reason relates to the closed nature of the markets in which they operate, due to the existence of special or exclusive rights granted by the Member States concerning the supply, provision or operation of the service concerned. Nowhere is this more the case than in the water and wastewater sector where the incumbent has typically an exclusive and monopoly right to provide services in a given geographical area.

In this section we detail, briefly, the main issues related to this Directive and the impact it has on the water and wastewater sector.

Equal treatment and transparency: In order to guarantee the opening up to competition of public procurement contracts awarded by entities operating in the water sector, it is advisable to draw up provisions for Community coordination of contracts above a certain value. Such coordination is based on the requirements inferable from Articles 14, 28 and 49 of the EC Treaty and from Article 97 of the Euratom Treaty, namely the principle of equal treatment, of which the principle of non-discrimination is fundamental. Additionally, the

---

[143] Directive 2004/17/EC of the European Parliament and of the Council of 31 March 2004.

[144] Directive 2004/18/ec of the European Parliament and of the Council of 31 march 2004 on the coordination of procedures for the award of public works contracts, public supply contracts and public service contracts.

principles of mutual recognition and proportionality, as well as transparency, must be respected. Accordingly, procurement needs to ensure that it follows these principles while providing sufficient flexibility. Similarly, the procedures developed should ensure equal treatment for contracting entities operating in the public and private sectors.

Single or multiple contracts: It must be noted that the Directive does not seek to prescribe whether the design and execution of works should be done separately or jointly. This decision should be determined by qualitative and economic criteria, which may be defined by national law.

Classification of contracts: the Directive provides for the classification of the different types of contracts for works as defined in Chapter 3. Service contracts may under certain circumstances include works contracts.

Tendering and procurement processes: The Directive provides for the free provision of services without discrimination. Procurement techniques to be used by the Member Countries should comply with the principles of equal treatment, non-discrimination and transparency. It is appropriate for the contracting entities to apply common procurement procedures in respect of their activities relating to water. However, procurement rules of the type proposed for supplies of goods are inappropriate for purchases of water, given the need to procure water from sources near the area in which it will be used. Qualification systems developed as part of the tendering system should be set with objective criteria, which, at the contracting entities' choice, may concern the capacities of the economic operators and/or the characteristics of the works, supplies or services covered by the system.

Special or exclusive rights: The fact that an entity supplies water under a special or exclusive right granted by a competent authority of a Member State, does not mean that the provisions of this Directive do not apply to the entity concerned. However, it is possible – subject to not introducing distortions in competition – to exclude certain service, supply and works contracts awarded to an affiliate undertaking having its principal activity the provision of such services, supply or works to the group of which it is part, rather than offering them on the market. It is also appropriate to exclude certain service, supply and works contracts awarded by a contracting entity to a joint venture which is formed by a number of contracting entities for the purpose of carrying out activities covered by this Directive and of which that entity is part.

Technical Specifications: technical specifications should allow public procurement to be opened up to competition and different technical solutions to be presented should be possible. The Directive allows for the inclusion of specific performance conditions provided they are not directly or indirectly limiting competition.

Mutual recognition: when evidence of a particular qualification is required as part of the tender process, the principles and rules of mutual recognition of evidence of qualification apply.

Award of contracts: is to be made on the basis of objective criteria which ensure the principles of transparency, non discrimination and equal treatment, and which guarantee that all parties to the tender are treated in conditions of effective competition. The determination of the award criteria depends on the object of the contract and the award conditions should fulfil economic and qualitative criteria.

5.2.2 Implications of public involvement & ownership under EU

Although our analysis is concerned with the introduction of the private sector, Community Law is not concerned with the mechanisms through which the private sector is involved, but rather the degree of control exercised by the public sector.[145] This is subject to Article 85(1), which ensures that the rules contained in the Treaty must be maintained. In a practical sense, what this means in the context of this book is that even though a Government may have an 'interest'[146] in a water utility, it must still abide by the basic principles of Community law and in particular the requirements of the Water Framework Directive.

At the same time, another important consideration in this book is the extent of public ownership in the sector. Actually, the Treaty does not prohibit public ownership, nor indeed does it limit it from enhancing its current area of service provision so long as it does not break any fundamental principles of Community law.[147] Thus, Article 222 of the Treaty states that the free market ethic that characterises the economic system of the EU will in no way affect the system of property ownership. For us, the implication of such a position is that, in principle, public undertakings are subject to the same rules of Competition law as the private sector.[148]

---

[145] *See* Cases 188–199/80, *France v Commission* (1982) ECR 2545 at 2579: Public undertakings are any undertakings over which the public authorities may exercise directly or indirectly dominant influence by virtue of their ownership of it, their financial participation therein, or the rules which govern it.

[146] *Id.,* Cases 188–199/80.

[147] *See* ECJ Decision 82/861, OJ 1982 L 360/36. The 1982 ruling by the Commission on British Telecom and its abuse of dominant position in relation to messaging services.

[148] *See* Case 155/73, *Sacchi v Commission* (1974) ECR 409 at 430. *See* also "Eleventh Report on Community Competition Policy (1980) at 155".

5.2.3 Implications of State Aid

The amount of state aid in any particular situation may vary and its impact has to be carefully considered. The first thing to consider is if the country itself has abided by the rules regarding state subsidy whose range is limited in Article 87(1)[149] of the Treaty.[150] However, there are a number of cases deemed to be compatible with the common market[151] but that should be carefully authorised by the Commission. These are set out in Article 87(2) and 87(3) of the Treaty.

Some of these are very important in the sector under consideration and those relevant issues should be noted, including: aid of a social nature granted to individual consumers provided there has been no discrimination,[152] aid aimed at promoting the economic development of specific areas,[153&154] aid seeking to promote the implementation of a major project of common European interest, or rectifying a serious disruption in a Member State.[155]

The particular case of the water sector, and notably in the contracts developed in all the area and Estonia which are further developed in this book, state aid was a definite condition under which both contracts with the private sector were obliged to operate. A government could not be seen to favour any one contractor through subsidies, either to businesses or consumers. Thus, the aid available had to be on a fair and transparent basis for all operators.

5.2.4 EU Competition Law

In the provision of public services such as water and wastewater it is essential to first ascertain whether or not EU Competition law is applicable and if so, in what context and to what extent.

A number of principles for determining whether a firm has a dominant position have been stipulated by the European Court of Justice in the application of case law. These principles can be used as *'tests'* to ascertain how a water

---

[149] Prior to Treaty of Amsterdam, Article 87 was numbered as Article 92.

[150] *See* Art 87(2), EC Treaty, *supra* note 104, which states that unless the Treaty specifically stipulates the contrary, aids granted by Member States, using any public money whatsoever, which actually or potentially distort competition by favouring certain firms or industrial activities, are illegal in as much as they affect intra-Community trade.

[151] Whish, R.P., Competition Law, (1994) 4th Edition, London, Butterworths at 222.

[152] *See* Art 87(2), EC Treaty, *supra* note 104.

[153] *Id.,* Art 87(3).

[154] *Id.,* Art 87.

[155] *Id.,* Art 87.

56   Public and Private Participation in the Water and Wastewater Sector

utility, and hence a PPP option may be judged to be compliant with Competition Law. These tests are:[156]

Test 1: The capacity of a firm to engage in actions that do not take into account their competitors, customers or suppliers.

In case law, the European Court of Justice has defined dominant position to be *"a position of economic strength enjoyed by an undertaking which enables it to prevent competition being maintained in the relevant market by giving it the power to behave, to an appreciable extent, independently of its competitors, customers, and ultimately of its consumers".*[157] This concept captures the potential imbalance of power between a water utility and its customer base, both consumers and government. As a natural monopoly in a geographical region, the water utility clearly has the power to exhibit this form of dominance and potentially abuse it. Hence an effective regulatory framework is required to mitigate the effect of this type of dominance through a series of monitors, controls and, as necessary, sanctions.

Test 2: The capacity to prevent other undertakings/firms from entering the market

The main barrier to other operators entering a given geographical market is the existence of a single water utility infrastructure system. It is likely to be economically unattractive to build a duplicate infrastructure system. If a water utility were able to control competitors from having access to that infrastructure, either through veto or excessive tariffs, then it would constitute a dominant position. Similarly if it were able to prevent a competitor building new infrastructure, it would be in a dominant position. Hence a regulatory framework is required to ensure fair competition. This could be managed through the contract letting process to avoid giving the private operator of an existing infrastructure an unfair commercial advantage. This could be through regulating mutual access to infrastructure, or a mechanism to maintain an equal playing field regarding the benefits of building and operating new infrastructure.

Test 3: The market share of the undertaking in question.

The natural monopoly of the water utility in both its geographical and product market effectively gives customers little, or no choice of the supplier from whom they may buy water and wastewater services. Hence, the water utility's market share will be greater than 70 to 80%. The European Court of

---

[156] *See* Cairns, *supra note* at 205–6.
[157] Case 27/76, *op cit,* at 207.

Justice generally regards a market share of greater than 70 to 80% as a dominant position.[158] Usually a water utility is in this position and thus clearly dominant and consequently may be in the position to abuse it. Therefore, the regulatory regime must ensure that this type of dominance is mitigated through a series of monitors, controls and sanctions.

Test 4: The economies of scale realised by the undertaking/firm

Again, as a natural monopoly, the economy realised by a water utility with its almost total market share gives rise to economies of scale, particularly in the case of distribution. Undoubtedly this could prove a major economic barrier to competition. Hence, a regulatory regime should ensure that there is no economic advantage to be gained from being the first private sector operator in a geographical area, and that a mechanism is in place to ensure that competition is encouraged, if only in future bid contests.

Test 5: The capital strength of the firm and its access to capital markets.

In the water sector, this type of dominance is only partially relevant to the argument around competitiveness. Indeed a large element of the benefit of using PPPs is that they exploit the private sector's ability to raise capital. There is however, a risk that with the growing importance of the 'Water Trans-nationals', the need for capital, with the subsequently dominant position of the 'Water Trans-national', may lead to a poor deal for the customer on an ongoing basis, and hence be anti-competitive.

Test 6: The ownership structure

After a water utility has introduced some form of private involvement, say through the selling of a strategic interest, the key issue is whether the public sector continues to exercise control over the water utility's activities.

5.2.5 Other considerations related to EU Competition Law

Although this concept is not directly applicable in our case studies, Article 82 of the Treaty is concerned with competition that may *'affect trade between Member States'*. However, the geographical market definition of a water utility and its historical constraint within national borders has meant that there is no cross-border trade in water services. Notwithstanding the current situation, it may be possible in the future for the same international operators in adjacent

---

[158] Druesne, G., Droit matériel et politiques de la Communauté européenne, (PUF Paris, 1991) (in French).

nations[159] to trade water between them to gain economies of scale, or alternatively merge their commercial operations into a regional hub larger than a single national operation.

5.3 EU LEGAL POSITION ON PPP

Member Countries, recent entrants to the EU as well as Accession Countries are relying more and more on different forms of PPP's to meet infrastructure investment needs.[160] The form of PPP varies substantially between Member States and depending on the sector. National law also clearly influences the type of PPP structure chosen. In the water and wastewater sector, PPPs have increasingly been used by Accession Countries as a strategy to help meet EU accession criteria and more specifically the EU Water Framework Directive.

The EU's definition of 'PPP' is a blanket term for agreements between public authorities and private industry. Such agreements may concern public infrastructure projects or public services and are categorised as:

a) Purely Contractual PPP which refers to a partnership based solely on contractual links between the different players. It covers a variety of set-ups where one or more tasks are assigned to the private partner, and which can include the design, funding, execution, renovation or exploitation of a work or service.

b) Institutionalized PPPs refer to the establishment of an entity held jointly by the public partner and the private partner. The joint entity has the mandate to deliver work or a service for the benefit of the public.

c) Cooperation agreements: these are usually between local authorities.[161]

---

[159] *See* WFD, Chapter II Section IV, *supra* note 8. This possibility may also be encouraged by the river basin management concept that is enshrined in the Water Framework Directive.

[160] See Report on the Commission Green Paper on public-private partnerships and Community law on public contracts and concessions (COM(2004)0327 – 2006/2043(INI)), Committee on the Internal Market and Consumer Protection, pg 8, "On closer inspection it is remarkable how greatly the development of PPPs varies across the European Union. A rough distinction may be made between three groups of Member States: a leading group, comprising the UK, France, Germany, Ireland and Italy, a mid-field and a group with little experience in this area. In view of their particular economic and political features the central and eastern European countries form a group of their own. In these countries PPPs are playing a particularly important role in creating a modern infrastructure."

[161] See ruling of 13 January 2005 in Case C-84/03 (Commission v Spain). This ruling has caused significant debate amongst local authorities in Europe as to whether and in what circumstances forms of cooperation between local authorities are subject to procurement law. The ruling says that cooperating local authorities cannot as a rule be excluded from procurement law by a national

European Legal framework and its implications on PPP   59

Recently, significant steps have been taken by the Commission to develop an approach to PPP. To this end the Commission adopted the Green Paper on Public-Private Partnerships and Community Law on Public Contracts and Concessions[162] as well as subsequent communications.[163] This paper was the result of a consultative process[164] and was one of the actions resulting from the European Initiative for Growth.[165] The aim of this Green Paper has been to determine the suitability and applicability of Community law on public contracts and concessions involving PPP. It sets out the rules that should be applied when taking a decision to entrust a given activity a third party (usually the private sector) under a PPP. It must be noted however, that these rules are downstream of the economic and organisational choice made by a local or national authority. The Commission makes no value judgment regarding the decision to use PPP and thus externalise the management of public services or not. This decision remains squarely within the competence of public authorities.

5.4 EU CONTRACTING OPTIONS FOR PPP

Despite the basic guidelines or distinctions on PPPs, Community law on public contracts and concessions do not prescribe whether or not a PPP arrangement should be used. A Member Country, and thus by implication Accession Countries, are free to provide a public service themselves or to entrust it to a third party. Moreover, under PPP arrangements, public partners are primarily national, regional or local authorities. There may also be public law bodies which have been created for the specific purpose to fulfil certain specific interests under State control. For the purposes of EU interpretation, the term "contracting body" or "contracting authorities" is used to designate all of these entities.[166]

---

provision. The law on contracts always applies when two legally distinct entities conclude a contract for pecuniary interest. The status of the entities is not at issue.

[162] Green Paper on Public-Private Partnerships and Community Law on Public Contracts and Concessions, COM(2004) 327 final Brussels, 30.4.2004

[163] Green Paper on Public-Private Partnerships and Community Law on Public Contracts and Concessions, COM(2005) 569 final, Brussels, 15.11.2005

[164] See Communication from the Commission to the Council and to the Parliament "Public finances in EMU 2003", published in the European Economy No 3/2003 (COM (2003) 283 final).

[165] Communication from the Commission "A European initiative for growth: Investing in networks and knowledge for growth and jobs", COM (2003) 690 final, 11 November 2003. This report was approved by the Brussels European Council on 12 December 2003.

[166] The term "contracting body" or "contracting authorities" is considered the same and falls within the meaning of Directives 92/50/EEC, 93/36/EEC, 93/37/EEC and 2004/18/EC and the

However, it must be noted that for selecting the partner in a PPP, the general principles of the EC Treaty apply, as do the current directives on public contracts. Under the EC Treaty all agreements in which a public authority awards a contract are subject to the general principles on freedom of establishment and freedom to provide services in Articles 43 to 49.[167] It follows that PPP agreements must satisfy the requirements of transparency, equal treatment, proportionality and mutual recognition. In addition, PPP agreements that count as public contracts are subject to the application of the public procurement directives. Some special provisions apply to building concessions, while service concessions are not at present covered by secondary law. On the implementation of PPPs the national laws apply and they must be in agreement with the provisions of the EC Treaty, which take precedence. Additionally, the Commission has introduced some degree of flexibility under procurement law. Notably, it defined, on the basis of the rules and principles derived from the Treaty and applicable secondary legislation, the outlines of the concept of concession in Community law and the obligations incumbent on the public authorities when selecting the economic operators to whom the concessions are to be granted.[168]

5.4.1 Implications of PPP, national law and concessions

In the absence of a uniform legal framework for PPPs at European level and in view of the heterogeneity of national laws the question arises, whether uniform provisions are needed to safeguard transparency and effective competition between market participants. The EU has decided against the creation of a separate legal regime for PPPs, but considers that there is a need for legislative initiatives in the areas of concessions for the provision of services, and institutionalised public-private partnerships (IPPPs). EU also considers that as a matter of principle, the law on public contracts needs to be applied whenever a private partner is to be selected.

---

contracting entities of the type "public authorities" and "public undertakings" within the meaning of Directives 93/38/EEC and 2004/17/EC.

[167] The rules on the internal market, including the rules and principles governing public contracts and concessions, apply to any economic activity, i.e. any activity which consists in providing services, goods, or carrying out works in a market, even if these services, goods or works are intended to provide a "public service", as defined by a Member State.

[168] Interpretative Communication on concessions under Community law, OJ C 121, 29 April 2000.

Community secondary legislation establishes that for any contract[169] for the execution of works or services between a contracting body and an operator, these are to be designated as a "public works or public services contracts". The concept of "concession" is defined as a contract of the same type as a public contract except for the fact that the consideration for the works to be carried out or the services to be provided consists either solely in the right to exploit the construction or service, or in this right together with payment.

Thus, in relation to all PPP contracts the following should be considered:

- PPP arrangements should emphasise fairness and transparency. Accordingly, transparency must be in evidence whenever public funds are involved, as well as ensuring that the PPP arrangement allows for the right of elected representatives to inspect agreements and documents. Furthermore, the EU takes the view that transparent rules should be applied for the award of public contracts to ensure effective competition and protection from corruption, in the citizen's interest;
- For those PPP arrangements or contracts which have already taken place, and in order to avoid legal uncertainty, transitional periods should be applied for existing contracts that have been concluded in good faith in accordance with national law;
- PPP arrangements should make adequate provisions to ensure protection of the citizen's rights to ensure secure safety, efficiency and quality standards;
- Member States are recommended to alleviate the task of the public sector by means of standard contracts and by stepping up the training of the decision-makers who have the task of selecting the private partners for PPPs;

Furthermore the EU has decided not to establish a European agency for PPPs, but welcomes other ways of sharing experience, such as the networking of national and regional PPP authorities.

PPPs are essentially a form of public contracting. Given the significant examples under case-law of the Court of Justice,[170] the EU has decided against

---

[169] The contract has to be for a pecuniary interest i.e. a contract where a consideration in the form of a fee.

[170] See for example Case C-324/98. Telaustria Judgment, the Court stated in this respect that "[the] obligation of transparency which is imposed on the contracting authority consists in ensuring, for the benefit of any potential tenderer, a degree of advertising sufficient to enable the services market to be opened up to competition and the impartiality of procurement procedures to be reviewed". See also ruling of 30 May 2002, Case C-358/00, *Deutsche Bibliothek,* ECR. I-4685.

developing blanket regulations for PPPs. In any event, the decision has been taken by the Commission that the legislative process should as a matter of principle go carefully, because PPPs are often very complex agreements whose attractiveness can only survive in a climate of flexible regulation.

As stated earlier in this chapter, in legal terms there needs to be a clear statement that, wherever a private partner is being selected, transparent procedures have been used for this selection. Transparency and fair competition between market participants should be used no matter what form of PPP is used.

5.4.2 Purely Contractual PPP

As mentioned above, EU considers two types of PPP arrangements, Contractual PPPs and Institutional PPPs. In the case of PPPs that are characterised as public contracts the main issue is the question of the award procedure.

With PPP structures, the competitive procedure to be used tends to be slightly more complex. In this regard, a new award procedure has appeared on the scene to join the familiar public, restricted and negotiated procedures for procurement above the threshold values. This procedure combines elements of the tender procedure with those of the negotiated procedure and takes precedence over the negotiated procedure. It must be noted that difficulties do arise in defining precisely what is meant by 'particularly complex' contracts. In response to this, the Commission has clarified that a particularly complex project is that in which the public contractor is objectively unable to specify the technical means with which their needs and objectives can be fulfilled, and/or is objectively unable to specify the legal and/or financial conditions of a project.

Although PPP projects by varying nature need not necessarily be more complex, each case must be assessed on its own merits. In any event, the Commission does allow for the competitive dialogue procedure to take preference for the award of contracts in the PPP sector as a principle because it combines the advantages of the restricted procedure with those of the negotiated procedure. Its flexible nature enables it to obtain optimum solutions under great competitive pressure. Although its disadvantages include its complexity and the extent of effort required from all involved, these are offset by its greater transparency

---

These principles are also applicable to other State acts entrusting an economic service to a third party, as for example the contracts excluded from the scope of the Directives owing to the fact that they have a value below the threshold values laid down in the secondary legislation (Order of the Court of 3 December 2001, Case C-59/00, *Vestergaard*, ECR. I-9505), or so-called non-priority services.

compared with the negotiated procedure. This reduces the risk of anti-competitive behaviour.

5.4.3 PPPs as concessions

The concept of "concession" is defined as a contract of the same type as a public contract except for the fact that the consideration for the works to be carried out or the services to be provided consists either solely in the right to exploit the construction or service, or in this right together with payment. The Commission takes the view that concessions should be of limited duration so that competitors are not excluded from competition for an unnecessarily long time. At the same time it recognises that under these types of PPPs, competitive dialogue is also the appropriate procurement procedure, as its flexibility is as a rule suited to the complexity of concessions, without jeopardising the principles of transparency, equal treatment and proportionality.

5.4.4 Institutionalised PPPs (IPPPs)

IPPPs are joint industrial creations combining private and public partners. The objective of such a PPP structure is to provide or supply a service on behalf of the public, typically in the field of public welfare. IPPPs may be created by the formation of a mixed-economy enterprise or company, but may also arise when a private company assumes control of what was a public concern, i.e. by changing the share ownership.

This raises the question whether the two processes, new formation or a change of ownership, are relevant from the point of view of procurement law. The Commission says that the private partner of such an enterprise must be selected for the tasks involved in a transparent way and without discrimination, and must be selected irrespective of the actual form of the contract under the procurement directives or the EC Treaty. The selection of a private partner in accordance with objective criteria will safeguard fair competition.

Case law under the judgments of the Court of Justice in the Teckal and Stadt Halle cases (C-107/98 and C-26/03) set the principle that the participation of the awarding authority in a mixed-economy company does not justify exemption from the principles of procurement law. What is crucial is to award the contract to a body with independent legal personality. Accordingly an exemption from procurement law is recognised when the awarding authority exercises a control over the mixed-economy enterprise similar to that control which it exercises over its own departments, and when the enterprise essentially acts for the public body. The two criteria must be cumulatively fulfilled to ensure that there is equivalence with internal entities of the contracting authority. The Stadt Halle case-law has further clarified the matter. The Court decided that public contracting authorities may award a contract to an undertaking that belongs to them without adhering to the procurement provisions only when they hold 100% of the undertaking's capital, in other words when there is not even a small private shareholding in the undertaking. Only then, says the Court, does the contracting authority exercise control as it would over its own departments.

In summary, the Commission considers it necessary, in view of the transparency requirement and the ban on discrimination, for procurement law to be applied when an IPPP is set up, in so far as the act of setting it up is combined with the assignment of a public contract, concession or other public tasks to a private undertaking. Additionally, this same principle would apply when an IPPP is established by the sale of shares in a public undertaking.

5.5 WATER FRAMEWORK DIRECTIVE – BACKGROUND & DEVELOPMENT

It was not until the early 1970s that the EU Member States started developing legislation[171] to reach a common approach to water policy.[172] Initially this approach concentrated on deterring pollution and only for certain types of water or by certain pollutants.[173] The legislation was not an all-encompassing solution aimed at developing a sustainable pollution policy.[174]

Throughout the 1980s and 1990s[175] the remit of pollution control has been gradually extended, culminating in the first draft of the Water Framework Directive (WFD) issued in February 1997.[176]

---

[171] Blöch, H., The European Union Water Framework Directive: taking European water policy into the next millennium. (1999). Water Science and Technology 40(10): 67–71.

[172] The term *Water policy* is used to mean water and wastewater policy covering all aspects of water abstraction, management, collection, treatment, distribution as well as wastewater collection, treatment and discharge of both domestic and industrial.

[173] Commission Proposal for a Council Directive establishing a framework for Community action in the area of water policy, COM (97) 49 final, 1997 O.J. (C 184) 20.

[174] A sustainable Pollution policy should focus on pollution prevention, control and should go hand in hand with an environmental policy. The Member Countries should develop and implement mechanisms for pollution prevention.

[175] *Introduction to the new EU Water Framework Directive*, http://europa.eu.int/comm/environment/water/water-framework/overview.html

[176] *See* Commission Proposal for a Council Directive establishing a framework for Community action in the field of water policy; *supra* note 174, at 5.

The main objective of the proposed WFD was to create a framework for all future actions by the EU on water policies,[177] but at the same time be observant of regional variations.[178]

Thus, through the WFD the Commission sought to integrate the views of all stakeholders[179] including local and national authorities, water users, water providers, industry, agriculture, consumers, environmentalists, non-Governmental Organisations (NGO's) and other bodies involved in the water and wastewater sectors. It also aimed at ensuring coherence of objectives and eradicating conflicting policies between the sectors e.g. water, wastewater and environment, as well as any conflicting policies and/or objectives between EU member countries.

Fundamental to this approach was the need to rationalise water management policies by applying the concept of *'river basin'* management system,[180] which all governments would be required to adopt. However, a great deal of debate between the differing bodies, i.e. European Commission, European Parliament and the European Council, delayed the WFD implementation.[181] It was not until September 2000 that finally the EU adopted the WFD that entered into force on the 22nd December 2000.

In its current form, the WFD focuses on establishing the right conditions to encourage efficient and effective water protection at the local level, by providing a common approach and common objectives, principles, definitions and basic measures to achieve a sustainable water policy.[182] Thus, 'sustainable use' ensures adequate supplies of water of good quality to be maintained for present and future generations.

---

[177] International Environment Reporter, *EU Parliament Invokes Conciliation Procedures for final text of WFD*, http://www.subscript.bna.com/SAMPLES/ier.nsf

[178] *See* International Environment Reporter *id., at 15* "…variability of freshwater resources exists throughout Europe, with annual average runoff (water resource per unit area) ranging from over 3000 mm in western Norway to 1000 mm over large areas of Eastern Europe and less than 50 mm in southern and central Spain."

[179] *See id.,* at 20.

[180] *See* more detailed discussion on this point, *infra,* at 57.

[181] *See* Introduction to the new EU Water Framework Directive, *supra* note 176. After the Commission Proposal for a Council Directive to establish a Framework for Community Action in the Field of Water Policy in 1997, there were delays in the implementation of the Water Framework until September 2000 when an agreement was reached through numerous discussion seminars and meetings.

[182] Rieu-Clarke, A., The EU Water Framework Directive: A Model for sustainable use of water resources? University of Dundee, 2002.

It must be noted however, that the WFD recognises that its objectives are ambitious and that Member States will not be able to comply with them immediately.[183] It allows for extensions where improvements cannot reasonably be achieved.

5.5.1 Brief analysis of the WFD

The scope of the WFD covers all surface, ground transitional and coastal waters. Under Article 3[184] Member States must identify River Basin Districts[185] and a competent authority to manage them. The WFD also recognises that the management of these districts may imply regional, national and international co-operation and may also involve co-operation with non-Member States. It also explicitly permits existing structures stemming from international agreements to be used for this purpose.[186]

Thus, it can be appreciated from the previous section on the history and development of the WFD, that it has two equally important objectives. A *managerial* objective, aimed at creating integrated water management at the river basin level to ensure overall co-ordination of water policy in the EU,[187] and an *environmental* objective, based on the premise of preventing further deterioration and achieving water quality of "good status". Good status means that it is not only about preventing deterioration but, if deterioration has already occurred, there is an obligation to restore water quality.

---

[183] *Id.*, at 6: Changing the way the water environment is managed cannot be achieved overnight. This is recognised by the forward-looking nature of the Directive timetable as outlined below: **2003** define River Basin Districts and appoint Competent Authorities (such as the Environment Agency in England and Wales) to carry out the work; **2004** assess the River Basins; **2006** start monitoring programmes; **2007** state issues and objectives for River Basins; **2008** consult on draft River Basin Management Plans; **2009 to 2012** enact the plans; **2013 to 2015** review the plans, and in **2015** meet Directive objectives.

[184] *See* WFD Article 3(7), *supra* note 8 at 7.

[185] *Id.*, Art 2(15) Definitions at 5, "River basin district" means the area of land and sea, made up of one or more neighbouring river basins together with their associated ground waters and coastal waters, which is identified under Article 3(1) as the main unit for management of river basins.

[186] *Id.*, Art 3 at 7 item 4 "For international river basin districts the Member States concerned shall together ensure this coordination and may, for this purpose, use existing structures stemming from international agreements."

[187] European Freshwater Programme, 28 May 2001, http://www.panda.org/europe/freshwater/initiatives/wfd.html#good status

5.5.2 Summary of environmental objectives of the WFD

The parameters of the WFD, as specified in Article 1,[188] are to protect, enhance and prevent further deterioration of the aquatic ecosystems, promote sustainable water use based on long-term protection of available water resources, enhance protection and improvement of the environment, ensure the progressive reduction and prevention of pollution of groundwater, and to mitigate the effects of floods and droughts.

In order to do this, a sustainable environmental objective needs to be ensured, which will prevent deterioration and achieve *'good status'* in all waters by 2015 i.e. 'good surface waters' meeting certain ecological and chemical criteria, and 'good ground waters' meeting certain quantitative and chemical criteria). Simply stated, "status" means when the body of water in question deviates from its original or referential conditions. The objective of the WFD in this context is to set both quantitative and qualitative criteria[189] to be achieved by the

---

[188] *See* WFD Article 1, *supra* note 8 at 5–6.
Purpose: The purpose of this Directive is to establish a framework for the protection of inland surface waters, transitional waters, coastal waters and groundwater which:
  (a) prevents further deterioration and protects and enhances the status of aquatic ecosystems and, with regard to their water needs, terrestrial ecosystems and wetlands directly depending on the aquatic ecosystems;
  (b) promotes sustainable water use based on a long-term protection of available water resources;
  (c) aims at enhanced protection and improvement of the aquatic environment, inter alia, through specific measures for the progressive reduction of discharges, emissions and losses of priority substances and the cessation or phasing-out of discharges, emissions and losses of the priority hazardous substances;
  (d) ensures the progressive reduction of pollution of groundwater and prevents its further pollution, and
  (e) contributes to mitigating the effects of floods and droughts and thereby contributes to:
the provision of the sufficient supply of good quality surface water and groundwater as needed for sustainable, balanced and equitable water use, a significant reduction in pollution of groundwater, the protection of territorial and marine waters, and achieving the objectives of relevant international agreements, including those which aim to prevent and eliminate pollution of the marine environment, by Community action under Article 16(3) to cease or phase out discharges, emissions and losses of priority hazardous substances, with the ultimate aim of achieving concentrations in the marine environment near background values for naturally occurring substances and close to zero for man-made synthetic substances.

[189] *Id.*, Art 2 Definitions.
  17. "Surface water status" describes the status of a body of surface water, determined by its poorest ecological and chemical status factor.
  18. "Good surface water status" means that the status of a body of surface water has achieved at least "good" ecological and chemical status.
  19. "Groundwater status" describes the status of a body of groundwater, determined by its poorest quantitative and chemical status factor.

Member States. These detailed criteria stipulated in Annex 5 of the WFD are the basis for measuring sustainable use of waters.

The WFD is pragmatic in its approach to achieving its environmental objectives. It allows for *extensions* where improvements cannot reasonably be achieved. For example, exceptions may be granted when it would not be feasible, or would be disproportionately expensive (Article 4(5)). Other exceptions might include a temporary deterioration due to unforeseen natural circumstances (Article 4(6)).

Needless to say, the WFD is very specific about *water quality* standards. Accordingly, under Article 7, Member States are required to identify water currently in use or to be used in the future for human consumption, and to take all necessary protective measures to ensure the quality of these waters. Member States are also required to comply with existing and future legislative requirements. This necessitates measures to combat *pollution* whilst protecting the environment, which is a key principle of the WFD. Article 10 establishes emission limits and environmental quality standards. The WFD also focuses on the need to co-ordinate existing and future legislative requirements.

5.5.3 Summary of Managerial objectives of the WFD

Each River Basin must have river basin management plans[190] that include specific responsibilities to develop and set up appropriate administrative arrangements within each river basin, including the identification, or creation, of a competent authority to manage each River Basin District (RBD). In each plan, the characteristics of the river basin should be defined, a summary of the significant pressures on that particular basin identified, and a programme of appropriate measures to be developed.

In the case of international RBD's, each Member Country has to set up the relevant administrative arrangements within the portion of international RBD lying within its territory. In addition, each River Basin should design and

---

20. "Good groundwater status" means that a body of groundwater has achieved at least "good" quantitative and chemical status.
28. "Good quantitative status" is the status defined in table 2.1.2 of Annex V.
28. "Good quantitative status" is the status defined in table 2.1.2 of Annex V.
28. "Good quantitative status" is the status defined in table 2.1.2 of Annex V.
28. "Good quantitative status" is the status defined in table 2.1.2 of Annex V.
28. "Good quantitative status" is the status defined in table 2.1.2 of Annex V.
28. "Good quantitative status" is the status defined in table 2.1.2 of Annex V.
[190] *See* WFD Article 13(1) (2) (3), *supra* note 8.

European Legal framework and its implications on PPP 69

implement measures that will ensure that the objectives of the WFD are achieved, e.g. pollution control measures, water abstraction control, implementation of existing EU legislation for the protection of water[191] etc.

For international river basins that lie between Member States and Non-Member States, Member States "shall endeavour to establish appropriate coordination with the relevant non-Member States, with the aim of achieving the objectives of this Directive throughout the river basin district".[192] However, it is up to the Member State to apply the rules of the WFD within its respective territory.

Although the WFD states that "water is not a commercial product like any other",[193] there are some basic economic principles that the WFD endorses. In particular it states that Member States should take into account the *principle of full cost recovery* and the 'polluter pays' principle as provided for in Article 9. It should also take note of the economic analysis, the long-term forecast of supply and demand, volume, price and cost of water services, as well as the relevant investments for the provision of these services.

The WFD is specific about *water pricing policies* to be attained by 2010 to allow for adequate incentives for the efficient use of water. In addition, the WFD makes a distinction between the contributions that industry, agriculture and households provide in order to recover the cost of water services.

In the compliance and implementation of these WFD criteria, Article 8 provides for a *monitoring* structure. In it, each Member State must establish a monitoring programme by the year 2006. Ecological and chemical status for surface waters will be monitored, as well as the chemical and quantitative status of groundwater. Such monitoring should be in accordance with Annex V. The aim of these provisions is so that each Member Country may perform different types of monitoring and inform the European Commission of these arrangements in accordance with certain reporting mechanisms.

These *reporting* mechanisms provided for in Article 15 require each Member State to send copies of the river basin management plan to the Commission, the reports on the characteristics of river basins and the monitoring programmes that are being implemented, or proposed (Article 8). A further obligation is to submit an interim report describing progress to date and the

---

[191] *See* Introduction to the new EU Water Directive, *supra* note 176, at 21: There are 11 EU directives officially related to water protection and their correct application is vital for the good functioning of the WFD. However, some of these – Nitrates, Habitats, environmental impact assessment etc. – are some of the most un-implemented directives in the EU, with most EU countries facing the European Court of Justice because they are not fulfilling their obligations.

[192] *See* WFD Article 3(5), *supra* note 8, at 7.

[193] *Id.*, Preamble (1), Water is not a commercial product like any other, but rather, a heritage that must be protected, defended and treated as such.

analysis required under Article 5[194] within three years of implementing the river basin management plan.

Public participation is seen as an essential part of the WFD. Article 14[195] actively encourages each Member State to ensure that all interested parties are involved, especially in the production, review and updating of the river basin management plans which should be made available to the public for comment at least three years prior to adoption. All information used for the development of the plans should be made public.

Finally, it should be noted that compliance and implementation of the WFD cannot be fulfilled if there are no instruments in place to enforce adequate *penalties* for a breach in the WFD provisions. In this respect, the WFD allows penalties to be imposed by Member States under Article 23.

5.5.4 Framework for analysis of WFD in Water & Wastewater Utilities

The main objectives of the WFD have been discussed above. In order to prevent further deterioration of water resources, the WFD provides the principles to establish the right conditions to encourage efficient and effective water protection, thus ensuring adequate water quality. Below we extract the key criteria from the WFD's main principles, for our methodological framework for analysis.[196]

WFD 1. Water Quality – "Good Status"
As described before, the issue of water quality and its good status should focus on preventing deterioration of 'status' in all waters and to achieve 'good status' (sustainable) in all waters by 2015 in:

- 'Good Surface Waters' within Ecological & Chemical Criteria
- 'Good Groundwater' within Quantitative & Chemical Criteria

---

[194] *Id.,* Art 5 – Characteristics of the river basin district, review of the environmental impact of human activity and economic analysis of water use.

1. Each Member State shall ensure that for each river basin district or for the portion of an international river basin district falling within its territory: an analysis of its characteristics, a review of the impact of human activity on the status of surface waters and on groundwater, an economic analysis of water use is undertaken according to the technical specifications set out in Annexes II and III and that it is completed at the latest four years after the date of entry into force of this Directive.

2. The analyses and reviews mentioned under paragraph 1 shall be reviewed, and if necessary updated at the latest 13 years after the date of entry into force of this Directive and every six years thereafter.

[195] *Id.,* Art 14(1) (a).
[196] *See* Annex 1.

In addition, we will assess the application of quantitative and qualitative criteria and the application of detailed criteria for measuring 'sustainable use'. We will also review whether there are any 'extensions' and when necessary improvements 'cannot reasonably be achieved'. We will also look at the possible exceptions:

- Art 4(5) non feasible or disproportionately expensive
- Art 4(6) temporary deterioration due to unforeseen natural circumstances
- Art 4(7) results in modifications of waters and benefiting human health, safety or sustainable development.

WFD 2. Management by River Basin
Management by river basin should be set out in River Basin Management Plans (Art 13). This indicator will look at the plans and assess whether or not to include characteristics of River Basin, a summary of significant pressures and human activity and a programme of measures. In addition, management of International River Basins should also be reviewed and in particular with respect to international management of river basins within EU (i.e. member countries must produce a plan) and co-ordinated with others in the overall plan. For those river basins that extend beyond EU, countries should endeavour to produce a single plan, and at the very least plan for Member State.

WFD 3. Public participation
Public participation is one of the criteria which are considered essential in our framework for assessment, because the services that are provided are fundamental for public well being. Thus, we will seek to ensure public participation and consultation (as provided for in Article 14 of the WFD). We will review whether this has been achieved and whether the parties were available for comments. In addition, there should be a timetable and work programme for public participation (3 yrs prior to adoption), an interim overview of significant water management issues, draft copies of River Basin Management Plan, and we will check if there is public access to information and background documents.

WFD 4. Monitoring and Reporting
Monitoring and reporting is provided for in Article 8 of the WFD. We will therefore ensure that there is an established monitoring programme (by the end of year 2006) which should include surface water (to cover issues such as volume and level, or rate of flow; ecological and chemical status; and ecological potential), ground water (to cover issues as chemical and quantitative status, technical specifications and standardised methods).

In addition, in this indicator we will look at the available Reporting regimes as provided for in Article 15. This should include as a minimum a summary of reports of monitoring programmes implemented (Art 8), review of impact of human activity on status of waters (Art 5) and an economic analysis of water use (Art 5).

WFD 5. Cost recovery

This indicator is basic, as the involvement of the private sector is based on economic principles. Thus we will review how the principle of full cost recovery has been applied, in particular covering issues such as an economic analysis that covers long-term forecast of supply and demand, volume, price and costs of water services, and the relevant investments. In addition we will assess whether or not the 'polluter pays' principle is being applied. We will also review the water pricing policies which should provide adequate incentives for efficient water use by 2010 and must ensure that there is an adequate contribution to recover costs of water services from different water users (including industry, agriculture and households).

6
Framework to achieve consistency between a PPP structure and EU law

6.1 RATIONALE BEHIND THE PROPOSED FRAMEWORK TO ACHIEVE CONSISTENCY BETWEEN A PPP STRUCTURE AND EU LAW

The objective of this book is to consider whether the mechanisms available to Accession Countries by *'partnering'* with the private sector provide a sound legal basis to achieve compliance with Community law and related directives. The book looks at how the private sector has supported two different countries,

© 2009 IWA Publishing. *Public and Private Participation in the Water and Wastewater Sector: Developing Sustainable Legal Mechanisms*, by X Cledan Mandri-Perrott. ISBN: 9781789065299. Published by IWA Publishing, London, UK.

74   Public and Private Participation in the Water and Wastewater Sector

one in Sofia, Bulgaria and the other in Tallinn, Estonia,[197] in helping meet some of the accession criteria.

This chapter looks at setting out a practical methodology that will act as a framework to assess the manner in which a PPP structure may aid meeting certain accession criteria. One of the problems of defining a framework and empirical set of tools like the one proposed in this book relates to the current controversy about the use of PPP in the provision of water and wastewater.[198] The debate as to whether the private sector is motivated simply by 'profit'[199] should be considered. There is a generalized view that, given the right financial and economic instruments and incentives, the public sector is equally capable of achieving the same benefits as the private sector.[200] Certainly this may well be true in countries with a well developed economic structure, but one must remember that few countries applying for accession are in the privileged position of having 'the right financial and economic instruments and incentives'. Accession countries have to meet a large number of accession criteria. The suggestion is that the private sector may contribute significantly to meet these criteria by bringing to bear managerial, operational and financial resources.

The challenge however is ensuring that whatever form of PPP is developed complies with national law, is consistent with its national sector policies, and at the same time will not fall foul of EU law. Governments of accession

---

[197] Both Bulgaria and Estonia lived through the transition from a command economy to market driven economies after the break-up of the Soviet Union. Government owned companies managed the provision of water and wastewater services in the City of Sofia in Bulgaria and the City of Tallinn in Estonia. Both suffered, amongst other things, chronic under investment, subsidised tariffs, poor levels of customer service, pressure to meet higher standards in water quality and effluent discharge, poor billing and collection, resulting in a financially unviable utility, inefficient operation with high levels of unaccounted water.

Faced with these challenges and unable to raise the required investment necessary to meet standards, both Bulgaria and Estonia examined the possibility of introducing the private sector for the provision of water and wastewater in the cities of Sofia and Tallinn respectively.

[198] Yaron, G., *The Final Frontier: A Working Paper on the Big 10 Global Water Corporations*, (2000) Toronto, Polaris Institute Publishers, at 69: Powerful financial institutions serve to fuel the water privatisation process by providing capital for projects sponsored by the major water corporations. Such financial institutions have the clout to impose water privatisation on governments as a condition for making the loans for the projects. The World Bank recently adopted a policy of making water privatisation a condition for its loans to developing countries.

[199] See id., 8.

[200] Hall, D., *Water Privatisation: Global Domination by a few*, (2000) Toronto, Polaris Institute Publishers, at 4: In 1997 the World Bank arranged the privatisation of the water in Manila, Philippines, despite the Bank itself describing the public water authority as a mature organisation with well-qualified staff, led by an able administrator.

Framework to achieve consistency between a PPP structure and EU law 75

countries are faced with complex decisions influenced by multiple objectives, not least meeting accession criteria. Accordingly, in the following sections we offer a multi-criteria framework that specifically aims at ensuring that these often competing factors are brought together in a consistent and sustainable way.

In this respect, this chapter sets out a practical framework to aid this decision making process. It involves a summary of several key aspects of the various criteria considered fundamental in achieving EU accession. Thus, the framework for analysis consists of three main areas:

1. Framework 1 related to EU Competition Law seeks to determine, under the given form of PPP structure developed what is the extent of public involvement, what is the relevant market of the service provider, and if there is potential for an abuse of its dominant position. Furthermore, under the framework, we look at whether the regulatory mechanism in place (or being proposed as part of the PPP structure) supports the basic EU Competition principles.
2. Framework 2 looks at the regulatory characteristics of the form of PPP proposed. In particular it seeks to understand whether the proposed contractual and institutional structure provides for a good definition of roles and responsibilities between the parties, and whether there is an adequate alignment between the PPP contract and the Member County's national law, including the degree of flexibility of the proposed structure to deal with long-term change of circumstances. Ultimately it seeks to determine the level of independence, accountability and transparency of regulation within the Member Country.
3. Framework 3 looks specifically at the EU's Water framework Directive and how the proposed PPP structures deal with the different WFD principles.

Chapters 7 and 8 apply these three frameworks systematically to the PPP processes of Sofia-Bulgaria and Tallinn-Estonia.[201] These examples of Accession Countries[202] were chosen as they have contrasting characteristics and therefore

---

[201] Fontaine, P, Europe in Ten Points, (1998) European Commission N° cat. PD-13-98-257-EN-C, at 6: Between 1946 and 1989 both Bulgaria and Estonia were occupied by the Soviet Union. After the fall of the Soviet Union in 1989 there was a major change in the manner in which both had to be organised. After more than 40 years of a *'command economy'* the transition to a *'market driven'* economy is well under way. Fundamental to this transition was the commitment by the European Community re-unify the continent and to admit countries such as Bulgaria and Estonia.

[202] *See* The Economist, *supra* note 16.

different WFD applications. They have entered into long-term arrangements with the private sector as a means of ensuring that they can achieve accession to the European Union. In each example we explore what the conditions prior to PPP involvement were, how the process was designed, and we apply this multi-criteria framework to assess the manner in which accession principles is being met. Existing contractual documents have been used as the basis for analysis of the two case studies. All information utilised for the analysis is in the public domain or the author has been granted permission to use the material that is referenced in this study.

6.2 OVERVIEW OF THE PROPOSED FRAMEWORK FOR PPP COMPLIANCE WITH EU LAW

As mentioned above there are three frameworks which can assist in ensuring that the proposed PPP contract or agreement that is developed is consistent with EU law. Below we detail the characteristics which are examined under each framework.

6.2.1 Framework 1: EU Competition Law

The first framework is mainly linked to EU Competition Law. Accordingly, our first step is to examine the *extent of public involvement*. In particular we will look at (a) the role of the public and its involvement, (b) the structure of PPP chosen and rationale for choice, and (c) whether or not there has been an equal and transparent treatment of bidders for the PPP contract.

The second step in the application of the framework related to EU Competition Law is to examine is to determine if there is an abuse of ***Dominant Position***. In particular we seek to determine what monitors and controls are in place to protect consumers from abuse of dominant position. Controls over pricing also need to be examined in particular with respect to the tariff structure and the tariff review mechanism.

The concept of ***Relevant Market*** is tested through the definition of a geographical market. In this case, we look at how the proposed PPP agreement defines the market area.

Another test is the determination of ***State Aid*** and whether there is any, and if so did it cause any distortions to competition at the time of bidding for the proposed PPP structure.

The last test in this framework is to determine whether there is a Regulatory mechanism to support ***Competition Principles***. In this, we seek to evaluate what mechanism is available to accommodate changes in national government policy whilst complying with European law. This is important as under European law

Framework to achieve consistency between a PPP structure and EU law 77

a contract cannot limit the capacity of a sovereign government to issue new laws, regulations and decrees. Furthermore, we need to evaluate the mechanisms to allow eventual competition within the PPP structure being proposed.

6.2.2 Framework 2: Regulatory Provisions

The first test within this framework relates to *Predictability and adequate definition of the PPP Contract.* The second test in this framework is to assess the *alignment of the PPP contract with national law*. Thirdly, we test for *flexibility/adaptability to deal with long contract periods. Independence, accountability and transparency* comprise the fourth test.

6.2.3 Framework 3: Water Framework Directive

The first test for the WFD is *Water Quality*.[203] In particular the main issues of concern are related to the prevention of the deterioration of 'status' in all waters, to cover Quantitative & Chemical Criteria. This test will also seek to determine if there are any quantitative and qualitative criteria being used, criteria for measuring 'sustainable use', and whether or not extensions and exceptions should be granted.

The second test seeks to determine *Management by River Basin*.[204] This test seeks to also identify any *'special characteristics'*. The administrative arrangements within each river basin are also key, as well as understanding the main the responsibilities within the river basin district.

---

[203] WFD in relation to Water Quality looks at the deterioration of 'status' in all waters, achieving 'good status' (sustainable) in all waters by 2015 in 'Good Surface Waters' to cover Ecological & Chemical Criteria, and 'Good Groundwater' to cover Quantitative & Chemical Criteria. Essential is the quantitative and qualitative criteria being used, criteria for measuring 'sustainable use' and whether or not extensions and exceptions should be granted. Exceptions can only be granted as provided for in the WFD: Art4(5) infeasible, or disproportionately expensive; Art 4(6) temporary deterioration due to unforeseen natural circumstances; and Art 4(7) results of modifications to waters benefit human health, safety or sustainable development.

[204] River Basin Management Plans (Art 13) to include:
Characteristics of River Basin
Summary of significant pressures and human activity
Programme of measures
In addition, International River Basins within EU must co-ordinate with the aim of producing a single plan, and at least a plan for Member State, and for those extending beyond EU, they should endeavour to produce a single plan, and at least plan for Member State.

The third test is *public participation*. Under this test we seek to ensure that the proposed PPP arrangement established pro-active mechanisms to ensure that the public is informed and actively consulted.

The fourth test is whether or not there are adequate mechanisms for *Monitoring and Reporting*[205] as provided for in articles 8 and 15 of the WFD respectively. These include mechanisms with respect to quality and quantity of water and sources of the water, ecological impact, review of the impact of human activity, and economic analysis of water use.

The fifth and final test of the WFD is that of **Cost Recovery**. What this test does is to determine whether the principle of full cost recovery, including economic analysis, long-term forecast of supply and demand, volume, price and costs of water services, are being properly applied. In addition it looks at whether or not relevant investments are being made, determines if the "polluter pays" principle is being applied and whether water pricing policies provide adequate incentives for efficient water use. Finally it assesses if there is an adequate contribution from different water uses (including industry, agriculture and households) to recover costs of water services is being made.

The summary of all three frameworks is in Annex 2.

---

[205] Monitoring (Art 8)
- Establish monitoring programme (2006)
- Include:
  - Surface water: volume and level or rate of flow; ecological and chemical status; and ecological potential;
  - Ground water: chemical and quantitative status
  - Technical specifications and standardised methods

  Reporting: (Art 15)
- Summary of reports of implemented monitoring programmes (Art 8)
- Review of impact of human activity on status of waters (Art 5)
- Economic analysis of water use (Art 5)

7
The Sofia, Bulgaria Concession Contract

7.1 BULGARIA – HISTORICAL BACKGROUND TO PPP

The Republic of Bulgaria is a country in South-Eastern Europe. It has a population of 7.7m; GDP per capita is $6,200, growing at an estimated 5% per annum.[206] Thirty five percent of the population are estimated to be below the poverty line.[207] From 1992, in line with a wave of reforms being implemented by Central, Eastern and Baltic countries, the Bulgarian Government looked at private sector participation as a mechanism to drive change. However, at the end of 1996 and in early 1997, mass popular uprising forced the elected Communist Government from office and a period of economic instability and rampant

---

[206] The Economist Country Report: Bulgaria, 4 (2006), FT Publishing, London.
[207] *See id.*, 12.

© 2009 IWA Publishing. *Public and Private Participation in the Water and Wastewater Sector: Developing Sustainable Legal Mechanisms*, by X Cledan Mandri-Perrott. ISBN: 9781789065299. Published by IWA Publishing, London, UK.

inflation followed (Inflation was over 1000% in 1997). The economic reasons for the 1996/97 crash were principally due to inconsistency and slow pace of reform, potential of reform achievements and the actual reverse, state control of production retained, substantial price distortions, artificial economic expansion, state owned enterprises run at a loss but kept afloat by the banking system. In July 1997 the Government implemented its Currency Board and a series of economic austerity measures aimed at stabilising the national economy. This led to a period between 1998 and 1999 in which the level of Foreign Direct Investment increased to over US$400 million.

However, during the Balkan war Bulgaria's growth was significantly restricted to an average of 2.5% per annum. Despite political infighting and constant changes in government, through continued efforts to establish economic stability Bulgaria managed to develop a stable enough strategy to allow the private sector to become comfortable with investments. By the end of 1998 over 28.1% of the companies selected for some form of private sector participation had been successful.[208&209] *"This is when the start of the privatisation and private sector development process really began"*.[210]

The initial accession target for Bulgaria was 2006,[211] although actual accession was 2007. In support of the accession process a number mechanisms have been implemented by the Community.[212] Of particular relevance in the environmental sector is the ISPA programme, which finances infrastructure and transport programmes. Although part of the ISPA funds are concerned with

---

[208] *See* The Economist Country Report, *supra* note 207.

[209] Vorkink, A.N., CASE STUDY: Bulgaria. (1999), Part I, 4, World Bank Publications, Country Director Bulgaria and Romania Country Unit, ECSPF.

[210] Telephone interview with Mr Tim Lowe, Business Development Manager for International Water and Project Manager responsible for the Sofia Concession. (12 February 2002). International Water is now the concessionaire for the Sofia water and wastewater company after a competitive process started in 1999. Mr Lowe indicated that it was at this point when his company started to look seriously at developing business in the water and wastewater sector.

[211] European Commission, Regular Report from the Commission on Bulgaria's progress towards accession, 9 (2001), Brussels: Office for Official Publications of the European Communities, 13.11.2001, SEC 1744.

[212] *See id.,* 9: There are three pre-accession instruments financed by the European Community to assist applicant countries of Central and Eastern Europe in their pre-accession preparations: the Phare programme; SAPARD, which provides aid for agricultural and rural development; and ISPA, which finances infrastructure projects in the fields of environment and transport. These programmes concentrate their support on the Accession Partnership priorities that help the candidate countries to fulfil the membership criteria.

For the years 2002–2003, total indicative financial allocations for Bulgaria amount to 100 million Euro from Phare, 53 million Euro from SAPARD, and between 83 and 125 million Euro from ISPA annually.

infrastructure projects[213] related to the environment, its principal priorities are *"the water sector's drinking water supply and wastewater treatment, solid waste management and air pollution"*.[214]

Coupled with the ISPA strategy was the Government's political willingness to include the private sector in the provision of water and wastewater services in the country.

7.2 PRIVATE SECTOR PARTICIPATION IN SOFIA

The water and wastewater company in Sofia called Vodosnadjavane I Kanalizatsia (ViK) was established in 1884 and was fully owned by the Municipality of Sofia (MoS). It served the City of Sofia and its 1.2 million population. The key problems facing ViK prior to the introduction of the private sector were a historic pattern of low and misallocated capital investment, inefficient operation – due to the old water and wastewater network, very low levels of capital investment and inefficient management practices, especially in billing and collection of tariffs. Added to these, there was more than 60% of unaccounted water, as well as high infiltration into the sewerage network resulting in low customer satisfaction.[215]

The introduction of the private sector in the water and wastewater services was conditioned by a number of issues stemming mainly from a political willingness to implement PPP and the ability to do so from a legislative perspective. In the case of Sofia, the commercial and political centre of Bulgaria, a strong Municipal Council and Mayor provided the political drive for PPP involvement.[216] Legally, the ownership of water and wastewater providers came under the jurisdiction of a regional ministry (MRDPW)[217] and the Municipality of Sofia[218] owned 100% of the water and wastewater services. These two factors facilitated the introduction of the PPP structure.

---

[213] *See id.,* 12: Emphasis has been placed on completing the main trans-European corridors, development of border connections including the new Danube bridge, electrification of railways and the reconstruction of the Sofia Airport.

[214] *See id.,* 12.

[215] Telephone interview with Mr Andrei Delchev, who spoke on behalf of the Municipality of Sofia (23 February 2002).

[216] At the time of the PPP involvement the Mayor of the City of Sofia was Mr Stefan Sofianski.

[217] *See* Vorkink, *supra* note 210. Outside the capital of Sofia, ownership of the water and wastewater providers is in the hands of the Regional Ministry of Regional Development and Public Works.

[218] *See* The Economist Country Report, *supra* note 207. The Municipality of Sofia had overall responsibility for water and wastewater services, roads and transportation, housing, public services including district heating and some power services. In the water and wastewater services the Municipality of Sofia was 100% owner.

There were two key objectives motivating the decision to introduce the private sector. First to demonstrate that private involvement was possible, and second to *"utilise private sector expertise and capital and so achieve significantly improved levels of service through the introduction of modern management, operation and maintenance practices and the planning, financing and execution of the investment programmes required to rehabilitate, operate, maintain and extend the existing system"*.[219]

7.3 OVERVIEW OF PPP DEVELOPMENT & PROCESS

As owners of the water and wastewater company, the Municipality of Sofia was keen to attract private sector capital,[220] gain operational efficiency, free up municipal management time, and encourage foreign investment.[221]

Faced with the problems identified above, in 1996 and with the support of the European Bank for Reconstruction and Development[222] (EBRD), the Mayor of Sofia started developing a strategy for private sector participation or PPP. In order to manage and co-ordinate the PPP process the MoS, hired international advisors[223] and established a Tender Evaluation Commission of 11 members.[224]

The PPP form chosen was a Concession i.e. full private sector involvement in the provision of water services. Using the analytical framework developed earlier, a full assessment will be given of both the Concession process in Sofia and its applicability to this case study.[225]

---

[219] Pre-qualification Information Memorandum in relation to the Private Sector Participation to operate the Sofia water supply and wastewater system, currently operated by the municipally owned company, ViK Sofia, (1999), reproduced with kind permission of the Municipality of Sofia, [hereinafter the Information Memorandum].

[220] *See* The Economist Country Report, *supra* note 207, at 5: At the time there were hardly no developed capital markets in Bulgaria.

[221] *See id.,* 3: Bulgaria was lagging behind other CEE in attracting Direct Foreign Investment.

[222] European Bank for Reconstruction and Development, *Sofia Water System Concession, Bulgaria, Project Summary Document.* http://www.ebrd.com/english/opera/projects/psd/psd2000/ Sofia concession loan Publications_files\438sofia.html

[223] *See id.,* 2, PriceWaterhouse Coopers (project managers), Cameron McKenna (international legal counsel), Eurolex (local legal advisors) and Hyder International (international technical and engineering advisors).

[224] The Membership of the MoS Tender Committee consisted of members of the Municipality, relevant ministries and members of the public. The intention was to establish a transparent and 'a-political' committee that was beyond reprisal.

[225] The figure in Annex 4 shows a diagrammatic representation of the impact the form of PPP has. There are two basic routes for evaluation: one in which the private sector has full involvement and the one where the public sector is still involved in the provision of services. In the case of Sofia, because the form of PPP that was chosen was a Concession, it means that EU Competition principles

During the definition period of the tender,[226] there was limited information on the mitigation of risks relating to the condition of underground assets and its subsequent impact on capital investment required. That's, it was decided that the structure of the PPP would allow for assets considered a national strategic interest to be retained and the risk of operation and investment would be allocated to the private sector by transferring rights of operation and exploitation. This led to the selection of the *concession* type of PPP structure. The Concessionaire would be responsible for the operation and maintenance of the water supply and sewerage system during the Concession Period including the design, planning, financing and construction of required capital investments. The MoS retained ownership of all existing water and wastewater infrastructure assets during the Concession Period. Ownership of new infrastructure assets, constructed by the Concessionaire was also vested in the MoS. The Concession Company was given the right to use those assets, both existing and future, in accordance with its rights and obligations under the Concession Contract.

The arrangements between the existing shareholders in the Concession Company and ViK were regulated through a shareholders agreement and an agreed set of articles of association for the Concession Company.[227] The length of the Concession was proposed for an initial 25 years[228] extendable for a further 10 years.

Obligations under the Concession Contract were defined primarily in terms of service, or output specifications. Failure to achieve the required service levels

---

apply i.e. we can proceed to doing the test for competition. In the event that the water utility still retained some public involvement, then it would have been necessary to do another test to determine the level of involvement of the public sector and the impact it may have. This test was actually done in the case of Estonia which is examined in the next chapter.

[226] Telephone interview with Mr Thomas Maier. Head of Eastern Europe for the European Bank for Reconstruction and development. (4 March 2001). Mr Maier indicated that the EBRD had been working closely with the MoS to develop the proposed Concession since early 1995. The EBRD and the MoS jointly funded the Advisory Team's fees under the EU Phare programme. This is an example where EU accession instruments were utilised in support of the implementation of the PPP strategy.

[227] Sofia Concession Agreement, (23 December 1999), Municipality of Sofia and Sofijska Voda AD concession agreement in relation to the supply of water and wastewater services for the Municipality of Sofia, [hereinafter Concession Agreement].

[228] Telephone interview with Mr Andrei Delchev, on behalf of the Municipality of Sofia (4 March 2002). Mr Delchev stated that the length of the Concession was initially set by the MoS for an initial 15 years extendable for a further 10 years, in accordance with the requirements of the Municipal Property Act (S.G. No. 44/21.05.1996). However, this was subsequently changed as it was felt that the capital investment requirements of US$150 million was too high to be adequately recovered over 15 years, or that the tariff level required to sustain such investment (and provide the Private Operator with adequate returns) would have to be too high.

would result in the imposition of financial penalties on the international operator/Concessionaire. In order to meet his obligations the Concessionaire was required to take over responsibility[229] for the infrastructure assets at the start of the Concession Period, as well as to operate, maintain and develop the Sofia water supply and sewerage system for the duration of the Concession. In addition, the Concessionaire was required to design, finance, construct and commission all works required to rehabilitate, and upgrade the water supply system[230] to achieve the required service levels.[231]

The Concessionaire was also responsible for metering, billing and revenue collection. Customers were charged in accordance with an agreed tariff schedule. The Concessionaire had powers to disconnect customers for non-payment.

A significant programme of capital investment was required in both the water supply and wastewater systems to meet target service standards and the Concessionaire was required to invest at least US$150 million in capital works over the Concession Period.

The bidding process resulted in the award of the concession to International Water.[232] The key factors in the successful bid were: a high level of committed investment (US$150 million) over a pre-defined period of time, enhanced customer service, a defined schedule to meet Water Framework Directives by the international operator, retraining of staff, and the establishment of outsourced businesses.

7.4 APPLICATION OF FRAMEWORK 1: EU COMPETITION LAW

It must be noted that because the company responsible for the provision of water services was 100% privately owned there was no need to undertake a preliminary test under Competition Law Article 222.

---

[229] *See* Section 5: General Rights and Obligations, Concession Agreement, *supra* note 228.

[230] *See id.*, 36, Produce a Drainage Area Study and Technical Plan for the rehabilitation and upgrading of the sewerage system to achieve the required service level; and design, finance, construct and commission all works identified by the Drainage Area Study and Technical Plan.

[231] *See* Information Memorandum, *supra* note 220, Fundamental to the Concessionaire's obligations was to comply with all relevant water and wastewater legislation as well as specific actions enabling compliance with the WFD, particularly in reference to River Basin Management obligations.

[232] *See* Vorkink, *supra* note 210. International Water is a company formed by Edison (an energy company in Italy), Bechtel International (a US construction company) and United Utilities (a UK multi utility company).

C1. Extent of public Involvement	• Role of the public and its involvement ➤ Has an exclusive right granted ➤ Extent of public control and compliance with WFD principles ➤ Structure of PPP chosen and rationale for choice • A PPP contract permitting equality and transparency to all bidders

The first step in the application of the framework related to EU Competition Law is to examine the *extent of public involvement*. In particular we will look at (a) the role of the public and its involvement, (b) the structure of PPP chosen and rationale for choice, and (c) whether or not there has been an equal and transparent treatment of bidders for the PPP contract. Since the selected PPP structure was a *Concession*, an exclusive right was granted to the concessionaire. This meant that although ownership of the assets was maintained in the public sector, the current assets were vested in the private operator for the duration of the contract. New assets were to be transferred to the MoS. Since there are no public ownership rights within the concession company, the examination will be conducted from a private perspective.

The bid process used was designed in a way that made it fully transparent. This included a process in three stages starting with a pre-qualification identification and background check of the potential bidders' capacity to manage the contract. This initial pre-qualification round, commenced in April 1999 and completed in May 1999, was followed by a detailed bidding round (a period between June to October 1999),[233] and then a final review prior to awarding the Concession. *Final submissions* consisted of a two-envelope submission that included, in envelope one, the lowest combined tariff from the bidders, and a second envelope containing an irrevocable commitment to a minimum capital investment requirement of US$150 million, and detailed technical strategies in areas such as asset management, customer care etc.[234]

---

[233] *See id.*, The result of the pre-qualification round was that the following international private operators were short listed, namely Berlinerwasser Betriebe, Vivendi, Lyonnaise des Eaux, International Water.
It is interesting to note that during the bidding period, Vivendi acquired a strategic stake in Berlinerwasser Betriebe and therefore it was the joint company that submitted the bid.

[234] *See id.*, 7–8. A factor of significant impact during the tender process was the relatively short timetable. This resulted in a number of issues that could not be resolved satisfactorily prior to, and during, the bid process. As a result, the pragmatic way forward was to establish series of *'Conditions Precedent'* in the concession contract *which* both the MoS and the private operator had to fulfil. The effect of this was that, although the contract was signed in December 1999, it did not become

86 Public and Private Participation in the Water and Wastewater Sector

The second step in the application of the framework related to EU Competition Law is to determine if there is an abuse of **Dominant Position**.

C2. Abuse of Dominant Position	• Market share greater than 70–80% constitutes dominant position • Monitors and controls to protect consumers from abuse of dominant position. • Monitors and controls to protect the environment from abuse of dominant position • Mechanism to control changes in ownership or operation of private sector businesses involved.

In this case, the Concessionaire is in a dominant position as its share is greater than 80%. Monitors and controls to protect consumers from abuse of dominant position were in place. Controls over pricing were built into the tariff structure and the tariff review mechanism agreed at the outset of the contract and subsequently agreed with the regulator/CMU. The CMU will ensure that the tariff remains competitive in comparison to benchmark sites elsewhere in Europe. The CMU was also vested with customer advocacy functions to ensure that the Client be represented before the Concessionaire Operator. Notably, the tender focussed on Levels of Service since the main objective of the MoS was to ensure compliance with the WFD.[235] The high level of capital investment required to attain these objectives necessitated a tariff increase to sustain this process. As a consequence, a relatively modest increase in tariff was levied (Approx. 12% real increase)[236] in the initial five-year period of the Concession. Thereafter, negotiations for tariff increases would be made at five yearly intervals. During negotiations with the international private operators, the issue of tariff review and its mechanism took on considerable importance. Thus, the contract was structured in such a way as to allow for automatic escalations in the tariff review mechanisms due to inflation and against a series of pre-defined eligible events.[237]

---

effective (i.e. reach *'Financial Close'*) until the first quarter of 2000. 'Financial Close' in the concession contract is defined as the moment in which both parties in the contract have fulfilled all conditions precedent, to the satisfaction of the other party.

[235] *See* more detailed discussion on this point, *infra*, at 93.

[236] Telephone interview with Mr David Carr, Project Manager for the Sofia Concession Project on behalf of International Water. (14 January 2002). Mr Carr indicated that the tariff increase of 12% is in real terms, which means that it is in addition to any escalation directly related to inflation.

[237] For confidentiality reasons, the eligible events cannot be enumerated but include changes in law and other so called 'grantor variations'.

In relation to instruments to monitor and protect the environment, Bulgaria had already instituted legislation that ensured water quality and wastewater standards for discharge to surface waters. However, specific waivers for wastewater treatment are detailed in the specific contract standards.[238] In this regard, EBRD and the EU, through the Phare programme, agreed to support the private operator in specific wastewater investments. Finally, in relation to a mechanism to control changes in ownership, or operation of the private sector businesses involved, the Concession contract allows for changes in ownership structure whereby the PO would be subject to a right of veto from the MoS to ensure a *'suitably qualified substitute'*.

C3. Relevant Market	• Geographical market defined • Significance of activities within this defined market • Product market –any interchangeable substitutes

The concept of **Relevant Market** is tested through the definition of a geographical market. In this case, the Concession Agreement defines the market area. However, as the Concessionaire becomes the exclusive operator in a given geographic area his activities are considered significant within the whole context of the water and wastewater sector. At the same time, as the product market is water and wastewater services there are no interchangeable substitutes.

Another test is the determination of **State Aid** and whether there is any, and if so did it cause any distortions to competition at the time of bidding.

C4. State Aid	• Any state aid that distorts competition when bidding
	• Exceptions:
	   ➢ Aid for social development
	   ➢ Aid for economic development
	   ➢ Promoting major European interest

At the time of the bidding/tender process neither the Government of Bulgaria, nor the Government of Sofia influenced the competition. However, there were some exceptions, as State Aid was included: (a) aid for social development was granted through a compromise between the Concessionaire and the MoS to ensure that the existing workforce would be transferred to the private operator and that no changes in numbers would occur during the first year of the concession. Thereafter, any changes in workforce would be at the sole discretion

---

[238] *See* Change in Control, Concession Agreement, *supra* note 228.

of the Concessionaire; (b) aid granted for economic development to achieve compliance with the EU standards for wastewater treatment – these, although in the operational scope of the Concessionaire, were not his responsibility to finance;[239] and (c) as the underground asset condition was unknown, rather than passing on the whole risk of the underground asset to the Concessionaire, a compromise was reached whereby the Concessionaire would develop an underground asset management plan within year two of the term, upon which, further capital investment programmes would be designed and mechanism to fund these developed.

The last test in this framework is to determine whether there is a Regulatory mechanism to support the *Competition Principles*.

C5. Regulatory mechanism to support Competition principles	• Mechanism to prevent other undertakings from entering market • Mechanism to accommodate changes in Community law or *acquis communautaire*. • Mechanism to control changes in national government policy within European law. • Mechanism to ensure eventual competition through limited duration contracts, or mechanisms encouraging competitors to enter the market at the same level as incumbent operator.

As the Concession Agreement granted an exclusive right to the Concessionaire to utilise the resources and assets in the given concession area there was in fact capacity to prevent other undertakings/firms from entering the market.

Is there a mechanism to accommodate changes in Community law or *acquis communautaire?* The chosen PPP structure envisaged that the relationship between the parties would be governed by a contractual mechanism defined in the Concession Contract. This contract detailed the obligations in terms of operations, maintenance, billing and collection, capital investment and other issues pertaining to the running of the water and wastewater services for the 'concession area'. Any changes in Community law were dealt within the contract through a *'Qualifying change of law'*[240] provision. These changes involve changes to the original conditions under which the Concessionaire undertook to provide the services due to new legislation.

---

[239] European Bank for Reconstruction and Development, *Sofia Bulgaria Funding, 438sofia.doc PSD* (First issued 7 Sept. 2000). File provided by the Bulgaria Project Desk of EBRD.

[240] *See* Performance Requirement Variations, Concession Agreement, *supra* note 228.

Is there a mechanism to control changes in national government policy whilst complying with European law? Under European law a contract cannot limit the capacity of a sovereign government to issue new laws, regulations and decrees. However, whilst this right of Government is not challenged, the contract does make a caveat for any changes in law that may affect the financial capacity of the concessionaire to meet its obligations. The Concession contract allows for changes in national law through *'Grantor Variation'* and includes a link to the tariff review mechanism, should this be required.

In relation to mechanisms to allow eventual competition the following must be noted. Because of the nature of the selected PPP, it meant that there was no direct competition for services for the successful bidder, in as much as the Concession Agreement grants an 'exclusive' right to operate within the City of Sofia and environs. However, this lack of competition was mitigated by the negotiation of tariff levels every five years through a clearly specified regulatory mechanism envisaged under the Contract's Concession Monitoring Unit.

7.5 APPLICATION OF FRAMEWORK 2: REGULATORY PROVISIONS

From the commencement of the design of the PPP strategy, it was recognised that there could be significant mitigation in the project's risk profile if the regulatory framework were to be fully developed. Unfortunately this was not possible as the legislation to comply with the *acquis communautaire* was still being enacted.

The first test within this framework relates to **Predictability and adequate definition of the PPP Contract**.

R1. Predictability/well defined PPP contract	• Structure used to assure sustainability • Current contractual structures • Adequate allocation of roles, rights, responsibilities and obligations between parties.

To avoid delays in the implementation of the PPP structure, the MoS decided that a *'regulatory function'* should be embedded into the Concession contract. This became known as the Concession Monitoring Unit (CMU) and was the structure that was used to ensure sustainability in the short and medium term. The key objective of this body was to undertake the monitoring functions of a 'traditional' regulator. Therefore, the Concession Contract granted the CMU certain rights, responsibilities and obligations vested in it by the MoS. In order for the CMU to be an effective regulatory tool, and not to be forgotten once the contract became effective, the establishment of the CMU was made a Condition Precedent to contract effectiveness. Thus, both the Concessionaire and the MoS had to agree on the scope and functions of the CMU according to the principles stipulated in the Concession Contract.[241]

The second test in this framework is to assess the *alignment of the PPP contract with national law*. In this case, national law for privatisation was first enacted in 1998. This was used as the basis for the development of specific water laws to establish the broad regulatory principles. However, these were incomplete at the time of the bid and therefore the CMU was created within the Concession Agreement.

R2. Adequate alignment of the PPP contract with national law	• Establish broad principles of regulatory policy using specific laws of the general and national water-sector. • Law to set national service standards. • PPP contract complimentary to such national law and covering specific commercial details of area served • Establish predictability and enforcement of the PPP contract, i.e. safeguards incorporated into the contract particularly relating to protecting public interest. • Any national legislation enabling alignment of accession criteria

In addition, national law already included quantitative and qualitative standards in line with the WFD. The Concession Agreement was in line with national law and covers specific commercial details related to water operations and the area served.

Predictability and enforcement of the regulatory functions were established in the Concession Agreement. It is very specific about the fines and penalties to be applied in the event of non-compliance.[242]

---

[241] Booz Allen Hamilton & Severn Trent Water International, *Draft Report on the Concession Monitoring Unit, Sofia Concession Contract*, (July 2000), at 5: The CMU is to be established by the Municipality of Sofia for the purposes of monitoring the compliance by the Concessionaire with the provisions of the Concession Agreement. The scope of authority of the CMU and the tasks to be undertaken by it, are to be agreed upon by the parties prior to the Effective Date and taking into consideration the provisions of Clause 30 of the Concession Agreement.

[242] *See* Penalties, Concession Agreement, *supra* note 228.

The national legislation enabling privatisation was the law enacted in 1992.[243]

Thirdly, we test for *Flexibility/adaptability to deal with long contract periods*. Flexibility has been achieved through the introduction of the CMU as a means of enabling regulation. See discussion on the fourth test below.

R3. Degree of flexibility/ adaptability to deal with long-term contracts	• Enable flexibility for adaptation and evolution for reaction in changing circumstances • Ensure no attempt of unfair advantage taken by any party should changes be required in the contract • Impartial supervision to ensure the original duties and obligations

The contract protects against unfair advantages through a precise mechanism for variations[244] in the Concession Agreement. This specifies when changes can be made and the procedure for such changes. The Concession also allows for a Dispute Resolution Procedure.[245]

Impartiality to ensure original obligations is achieved through the CMU which has a clear mandate, of the roles, responsibilities and obligations to which both the public and private sector subscribe and will abide by.

Independence, accountability and transparency are the fourth test.

R4. Independence, accountability and transparency of regulation	• Necessity of an independent and fully competent regulatory body • Regulator to uphold public and private interests • Ensure eventual competition • Ensures public participation

At present there is no national regulator but the parties expect that once the regulatory framework has been established, the functions of the CMU enshrined in the contract will be transferred to the national regulator. Thus, the CMU will either become the national regulator, or will be disbanded once its functions are covered by the national regulator.

---

[243] *See* Regular Report from the Commission on Bulgaria's progress towards accession, *supra* note 212, at 7: The privatization process in Bulgaria has suffered from instability of the legal framework – the 1992 Privatization Law has been amended 19 times.

[244] *See* Performance Requirement Variations, Concession Agreement, *supra* note 228.

[245] *Id.*, Dispute Resolution Procedure.

The CMU's role included customer advocacy and dispute resolution. Competition as such is not possible in the context of the Sofia contract. However, at a national level there is an intention to bring in other operators. Competition would thus be done through benchmarking and comparison of key indicators.

Public participation was ensured at the time of bidding through the special MoS consultative body in which representatives of customer groups were interviewed with their opinions included in the design of the PPP structure and, in particular, its indicators. In the future, public participation is ensured through the CMU's customer advocacy role that will eventually be migrated to the national regulator.

7.6 APPLICATION OF FRAMEWORK 3: WATER FRAMEWORK DIRECTIVE

The first test for the WFD is *Water Quality*. In particular the main issues of concern are the prevention of the deterioration of 'status' in all waters, achieving 'good status' (sustainable) in all waters by 2015 in 'Good Surface Waters' to cover Ecological & Chemical Criteria, and 'Good Groundwater' to cover Quantitative & Chemical Criteria. This test will also seek to determine if there are any quantitative and qualitative criteria being used, criteria for measuring 'sustainable use' and whether or not extensions and exceptions should be granted.

WFD 1. Water Quality – "Good Status"	Prevent deterioration of 'status' in all waters Achieve 'good status' (sustainable) in all waters by 2015 in: • 'Good Surface Waters' within Ecological & Chemical Criteria • 'Good Groundwater' within Quantitative & Chemical Criteria • Application of quantitative and qualitative criteria • Application of detailed criteria for measuring 'sustainable use' Extensions: - when necessary improvements 'cannot reasonably be achieved' Exceptions: Art 4(5) non feasible or disproportionately expensive Art 4(6) temporary deterioration due to unforeseen natural circumstances Art 4(7) results in modifications of waters and benefiting human health, safety or sustainable development.

In this regard, the Bulgarian government had already brought its legislation on water in line with EU legislation, including qualitative and quantitative criteria. However, the key focus in the Contract was the wastewater standards which, to achieve compliance, allowed a moratorium (in the form of a time extension after which these standards would be applicable), together with additional support from EBRD funding.

The contract needed to ensure that performance was monitored against the legislation and penalties applied for non-compliance. The CMU was established as a condition precedent to contract effectiveness in order to meet this principle.[246]

The second test seeks to determine *Management by River Basin.*

WFD 2. Management by River Basin	River Basin Management Plans (Art 13) to include: • Characteristics of River Basin • Summary of significant pressures and human activity • Programme of measures • International River Basins Within EU - must produce a plan for Member State and co-ordinate with others in an overall plan Extending beyond EU - should endeavour to produce single plan, and at least plan for Member State

In this case the Concession contract was designed to ensure that its outputs are reflective of and respect the WFD, e.g. pollution control measures, water abstraction control, implementation of existing EU legislation for the protection of water etc.

In addition, the contract was very specific in defining the areas of land together with their associated ground waters. The river basin district was also defined in the Concession, as was the Concession Area which complies with the WFD that requires the Concessionaire to identify its *'special characteristics'*.

The administrative arrangements within each river basin were the responsibility of the MoS and ultimately Government, including the creation of a River Basin District (RBD). The Concessionaire was not specifically responsible for river basin plans but does have an obligation to assist the relevant government authority in the preparation and production of these. This obligation is enshrined in the Concession Agreement.[247]

---

[246] Telephone interview with Mr Les Stachow. Head of PPP in Eastern Europe for Booz Allen&Hamilton, author of the Report on the Concession Monitoring Unit, Sofia Concession Contract, *supra* note 242, (15 August 2000). Mr Stachow said that the CMU did not have the power to impose penalties. The Municipality of Sofia would impose the penalties. The CMU provides a recommendation as to the relevance of the imposition of a penalty (i.e. should a penalty be applied) as well as a calculation of the amount of penalty imposed.

[247] *See* Annex 3 Levels of Service, Concession Agreement, *supra* note 228.

The third test is *public participation*

WFD 3. Public participation	Must ensure public participation (Art 14) and consultation, and be available for comments: • Plan timetable and work programme (3 yrs prior to adoption) • Interim overview of significant water management issues • Draft copies of River Basin Management Plan Public access to information and background documents

In this regard public participation was ensured at the time of bidding through the special MoS consultative body, in which representatives of customer groups were interviewed and their opinions included in the design of the PPP structure, particularly its indicators. Public participation is ensured in the future through the CMU's customer advocacy role that will eventually be migrated to the national regulator. Public access to information was granted to all CMU material presented by the Concessionaire.

The fourth test is whether or not there are adequate mechanisms for *Monitoring and Reporting*[248] as provided for in articles 8 and 15 of the WFD respectively.

WFD 4. Monitoring and Reporting	Monitoring (Art 8) Must establish monitoring programme (2006) Include: • Surface water: volume and level, or rate of flow; ecological and chemical status; and ecological potential • Ground water: chemical and quantitative status • Technical specifications and standardised methods

continued

---

[248] Monitoring (Art 8)
- Establish monitoring programme (2006)
- Include:
  ➢ Surface water: volume and level or rate of flow; ecological and chemical status; and ecological potential;
  ➢ Ground water: chemical and quantitative status
  ➢ Technical specifications and standardised methods
  Reporting: (Art 15)
- Summary of reports of implemented monitoring programmes (Art 8)
- Review of impact of human activity on status of waters (Art 5)
- Economic analysis of water use (Art 5)

	Reporting: (Art 15)
	Summary of reports of monitoring programmes implemented (Art 8)
	Review of impact of human activity on status of waters (Art 5)
	Economic analysis of water use (Art 5)

TheMoS had governmental agencies already in place to monitor drinking water quality and environmental issues.

Further controls needed to be instituted to protect service levels, tariffs to consumers and the specific WFD requirements in Annex 5. This was achieved through the contract itself[249] and the monitoring powers invested in the CMU.

The fifth and final test of the WFD is that of *Cost Recovery*.

| **WFD 5.** **Cost recovery** | Take into account the principle of full cost recovery
Follow an economic analysis
Long-term forecast of supply and demand
Volume, price and costs of water services
Relevant investments
In accordance with the 'polluter pays' principle
Water pricing policies must provide adequate incentives for efficient water use by 2010
Must be adequate contribution of different water users including industry, agriculture and households to recover costs of water services. |

What this test does is determine whether the principle of full cost recovery including economic analysis, long-term forecast of supply and demand, volume, price and costs of water services are being properly applied. In addition it looks at whether or not relevant investments are being made, determines if the 'polluter pays' principle is being applied and that water pricing policies provide adequate incentives for efficient water use. Finally it assesses if an adequate contribution from different water uses (including industry, agriculture and households) to recover costs of water services is being made.

---

[249] *See* Annex 5, Concession Agreement, *supra* note 228. The Concession Contract stipulated a series of very detailed reporting requirements from the Concessionaire. These included the main principles of the WFD Annex V e.g. to submit regular reports describing progress, including indicators, such as ecological and chemical status. In this the CMU had a very active role to play, as it was they who were to ensures compliance with this reporting mechanism as well as acting as a liaison with other government agencies e.g. Environment, River Basin District etc.

In response to this test, the proposed structure that was implemented looked at using a private partner to enable these principles to be fulfilled by the efficient management of the provision of services. The tariff mechanism developed in the Contract underpins this principle and avoids either direct or indirect subsidies. In addition the Concession Agreement included the 'polluter pays' principle.

Volumetric tariffs were applied, together with differing tariffs for various water users e.g. industry, agriculture and domestic.

7.7 BULGARIA SOFIA SUMMARY OF FINDINGS

The multi-criteria framework (EU Competition Law, Regulation and WFD) was systematically applied to the Sofia concession contract. Below we summarise how each of the issues of the framework have been dealt with by the Sofia concession contract:

1. **EU Competition law** principles:

Extent of public involvement was not achieved directly, in as much as the Concessionaire was granted unrestricted and exclusive use of assets, and new assets are to be the property of the Concessionaire. In addition, the MoS did not have an active shareholding in the Concessionaire Company.

Equal treatment of bidders was achieved as all bidders were granted full access to information and facilities, and no preferential treatment was given. EBRD played a key role as overseer of the bidding process to ensure transparency.

Given the monopoly rights that have been granted to the Concessionaire, the assessment for Dominant Position demonstrates that adequate *protection for customers* has been put in place. In addition *ownership* control mechanisms were established through the veto rights granted to the MoS in the Concession Agreement. However, *environmental* protection was only partly achieved, as the Agreement allowed for some specific waivers in wastewater discharge standards.

The test for relevant market requires an understanding of whether the company's activities are significant in a defined market area. Clearly, as the concessionaire has been granted an exclusive right to operate in that area there are no interchangeable substitutes.

State aid did exist related to storm water drainage and other specific issues where full responsibility was not transferred to the private sector.

2. With regard to **Regulatory principles**, the concession structure allowed:

Partial *independence and accountability* through the establishment of a regulatory function (by contract) in the form of a body called the Concession Monitoring Unit (CMU).

However, *competition promotion* was not achieved directly through the contract but this fact was mitigated in part through a competitive tariff structure (as the award criteria) and later on through the use of the CMU or national regulator.

Flexibility in regulatory functions is permitted in time, as the functions of the CMU are envisaged to be transferred to a national regulator once the necessary institutional structures are put in place.

3. WFD requirements:

With respect to *water quality* these have been partially fulfilled. The reason for this is that although water quality standards were in line with WFD, wastewater standards were not. The Concession Agreement allowed for a moratorium (in the form of a time extension of time after which these standards would be applicable) together with additional support from EBRD funding

The requirement for *management by river basin* is partly achieved through the concession structure, although the Concessionaire is not responsible for river basin management but takes an integral part through helping the relevant authority define plans and through its own use of the resource in the concession area.

Monitoring and reporting functions are well established in the Concession Agreement, as are the principles of *cost recovery* which form the basis of the tariff regime being applied.

It can thus be concluded that the form of PPP structure that has been chosen i.e. a concession, is an effective means to achieve compliance (even if not fully and immediately) with the three categories analysed. Again, its full effectiveness cannot be determined at this stage, but it does provide a good starting point for overall compliance. It can thus be argued that this type of PPP structure could assist other Accession Countries with similar conditions to meet the basic requirements of EU accession.

8
Tallinn, Estonia partial divestiture/sale contract

8.1 ESTONIA – HISTORICAL BACKGROUND TO PPP

The Republic of Estonia is a country in Eastern Europe, bordering the Baltic Sea and Gulf of Finland.[250] It has a population of 1.4m, GDP per capita of US$10,000 growing at 6%, the highest in Eastern Europe. In 1940 Estonia was forcibly annexed by the Soviet Union but was established as a separate and autonomous country after the break-up of the Soviet Union in 1991.[251]

---

[250] European Commission, Regular Report from the Commission on Estonia's progress towards accession, (2001), 35, Brussels, 13.11.2001: Office for Official Publications of the European Communities, SEC(2001) 1744.

[251] See id., 36.

© 2009 IWA Publishing. *Public and Private Participation in the Water and Wastewater Sector: Developing Sustainable Legal Mechanisms*, by X Cledan Mandri-Perrott. ISBN: 9781789065299. Published by IWA Publishing, London, UK.

Since then, there have been a number of economic strategies designed by Government to ease and support the involvement of the private sector or PPP.[252]

Successive Governments have *"focussed on reconstruction, deregulation and introduction of the private sector"*.[253] Due to inefficiencies in the old economic system and growing difficulties in exports to Russia, the Estonian economy experienced a severe recession in 1991 and 1992. GDP decreased by 12.6 per cent in 1991 and 14.2 per cent in 1992. The decline in GDP slowed from 1993, and by 1995 a real growth of the economy of approximately 4 per cent was achieved. A significant GDP growth was achieved in the three following years, especially in 1997, followed by a decline in real GDP of 1.1 per cent in 1999. The economic decline of 1999 can be attributed to the consequences of the currency devaluation in Russia in August of 1998 and the collapse of Eastern export markets.

Notwithstanding, Government has implemented a number of measures to ease and support Private Sector Participation, namely the currency reform of 1992 which basically replaced the Russian Rouble with the Estonian Kroon (EEK), which was then pegged to the German Mark at the rate of 8 EEK for 1 DM and finally, the implementation of a progressive tax policy e.g. not taxing profits until these are distributed and, dividends paid to Estonian residents not subject to further taxation. This has resulted in a successful track record of private sector involvement since 1992 in sectors such as manufacturing, district heating and banking.[254]

Estonia has been an 'accession candidate' for the EU since April 1997. Estonia has made *"considerable progress"*[255] in fulfilling the accession

---

[252] *Estonia Economic Report*, (2000), Suprema Securities, Estonia at 7.

[253] Telephone interview with Mr Ivar Virkus, Deputy Mayor of the City of Tallinn (12 December 2001). Mr Virkus explained his government's policy in relation to private sector participation in the water sector.

[254] *See* Regular Report from the Commission on Estonia's progress towards accession *supra* note 251 at 3.

The country has been a Member of the World Trade Organisation since November 1999 and is aspiring to NATO membership. The legal framework that underpins the involvement of the private sector is the Public Water Supply and Sewerage Act as well as the Water Act relating to specific City Council resolutions permitting the participation of the private sector in the water and wastewater company.

[255] *See id.,* 9: In its 1997 Opinion, the Commission concluded that Estonia fulfilled the political criteria. Since that time, the country has made considerable progress in further consolidating and deepening the stability of institutions guaranteeing democracy, the rule of law, human rights and respect for and protection of minorities. Over the past yea, further efforts have been made in this direction. Estonia continues to fulfil the Copenhagen criteria.

criteria. In addition, the Community and Estonia have implemented mechanisms to assist in this process.[256] ISPA funding in the environment sector was focussed on *"improvement of water management and wastewater treatment"*.[257] This has all resulted in Estonia gaining membership to the EU since May 2004.

8.2 PRIVATE SECTOR PARTICIPATION IN TALLINN

The company known as AS Tallinna Vesi (ASTV) is the largest provider of water distribution, wastewater treatment and storm water collection services in Tallinn, the capital of the Republic of Estonia. The Company serves the population of Tallinn of 415,300 and its suburb of Saue of 4,500.[258] In total, the Company has a customer base for drinking water of approximately 405,000. This is equivalent to approximately 99 per cent of the population.[259] Similarly,

---

[256] *See id.,* 9: Total financial assistance to Estonia for the years 2000–2002 amounted-annually to approximately 30 million Euros for **Phare,** 12.1 million Euros from **SAPARD**, and around 35 million Euros from **ISPA.**

[257] *See id.,* 12–13.
The Commission also mentioned: "In the environment sector, further progress has been made in relation to both alignment and implementation of environmental legislation. Estonia needs to continue these efforts and strengthen administrative capacity. Also enforcement of environmental legislation needs to be improved". However, there is still more to be done in relation to the water and wastewater sector, as was recently proven by the Commission's evaluation that concluded in its overall assessment that "…priority needs to be given to the complete transposition of the *acquis*, in particular as regards water quality and waste management".

[258] Information Memorandum, Tender Offer for Subscription of 30,000,000 A-Shares and Sale of 28,000,000 A-shares in AS Tallinna Vesi (Public Limited Company Incorporated Under the Laws of Estonia), (July 3rd 2000), [hereinafter Information Memorandum] reproduced by kind permission of AS Suprema Securities, Tallinn, Estonia, at 5: The Company supplied drinking water and sewage collection services to 15,911 clients as of January 1st 2000, including 3,598 industrial clients and 12,313 residential clients. Industrial clients of the Company include also apartment unions, established for collective management of residential properties that are charged a regular residential tariff.

[259] *See id.,* 6: The Company operates water distribution, sewerage and storm water networks. The lengths of these networks are, respectively, 820.5 kilometres, 739.4 kilometres and 151.0 kilometres. In the older parts of Tallinn, the sewerage network of the Company is combined with the surface water drainage network. It is this section of the Company that expects considerable expansion within the next six years, as it is the City of Tallinn's priority to fully develop the sewage collection system by 2006.

102   Public and Private Participation in the Water and Wastewater Sector

the Company's wastewater and sewage collection systems in Tallinn cover approximately 385,000 customers, which is approximately 95 per cent.[260&261]

The Government of the City of Tallinn (CoT) recognised the need to attract the private sector to the water and wastewater sector. It was an acknowledged fact that ASTV was a *"well run company"*[262] which could be attractive to the private sector. In addition, the City of Tallinn was under pressure to finance new social developments and to *'release'* itself from the administrative and financial burden of providing water and wastewater services. As ASTV was 100% owned by the CoT and not central government, there was no constraint in the decision to bring in a *"Private Operator/Strategic Investor"*[263] *(PO/SI)*.

The principal objective of the City of Tallinn was to *"attract a Strategic Investor for the management and financing of the Company in order to achieve the quality and service standards set by National and European Union regulations"*.[264]

8.3 OVERVIEW OF PPP DEVELOPMENT & PROCESS

The type of PPP structure selected was a partial divestiture through the sale of a majority stake in ASTV – the Company. This partnership between the public and private sector, through a divestiture or partial sale, was structured in such a way that the City of Tallinn (CoT) granted the Strategic Investor management and operational control, whilst the CoT retained a veto on a certain number

---

[260] *See id.,* 6: Tallinn Water and wastewater Company have net sales of US$ 27.5M (1997), 27.8 M (1998) and 24.6 M (1999) respectively.

[261] Telephone interview with Mr Chris Shugart, Senior Banker, European Bank for Reconstruction and Development (15 October 2001). Mr Shugart explained the role that the EBRD played with ASTV and the bank's commitment to private sector participation in the water sector. He mentioned that ASTV was considered one of EBRD's success stories being a recipient of US$ 20.5 M for capital investment projects from 1994 to 1999 prior to the introduction of the private sector. Mr Shugart mentioned that a good level of investment had already been made in underground assets, as well as water treatment technology through ozonation. He then provided a summary of what he felt were the key problems facing ASTV prior to the introduction of the private sector. These included inefficient operation due to an 'over dimensioned' water network resulting in poor quality stale water, underinvestment in wastewater network and an increasing risk of non-compliance with discharge standards into the Baltic Sea, billing inefficiencies and low collection rates, a need for improved technological, business management processes and procedures and a need to improve service levels to customers.

[262] Telephone interview with the Chairman of the Management Board and Director of Development, Mr Kalle Tiiter (16 October 2001).

[263] Resolution no. 210, City of Tallinn (June 15 2000) [hereinafter, City of Tallinn Resolution].

[264] *See* Information Memorandum, *supra* note 259.

of matters.[265] However, unlike the case in Sofia Bulgaria, ownership of the provider of water services was shared between the public and private sectors.

Due to this combined ownership structure, under EU Competition Law it was important to determine the extent of public involvement as provided for in Article 222. It follows that a preliminary test needed to be performed to gauge the level of control, exclusive right (Article 85(1)) and economic interest (Article 85(2)) before continuing with the competition framework used directly in the case of Sofia, Bulgaria.[266]

The process was arranged as a one-stage tender with pre-qualification. As a result of the bidding process, the strategic investor was to hold 58,000,000 A-shares of ASTV, representing 50.4 per cent of the outstanding share capital of ASTV after the issue of the above 30,000,000 new A-shares. This gave the Private Operator/Strategic Investor (PO/SI) management control of ASTV via majority in the Supervisory Board and the right to appoint members of the Company's Board of Management. The City of Tallinn, as the owner of the preferred B-share and party to the Shareholders' Agreement, retained control over key matters relating to the management of ASTV.[267]

The envisaged PPP structure consisted of a *Shareholders Agreement*[268] (SHA) between the City of Tallinn and the Private Operator/Strategic Investor[269] and a *Service Agreement*[270] defining the manner in which the Private Operator/Strategic

---

[265] *See id.,* at 4: The City of Tallinn's right of veto related mainly to the change in the PO/SI ownership structure and the selling of PO/SI's shares.

[266] *See* Annex 4.

[267] *See id.,* 9: Post-transaction Shareholder Structure:

Shareholder	Number of Shares	Nominal Value (EEK)	% Interest
City of Tallinn	57,000,000 A-shares 1 B-share	570,000,000 1,000	49.6
Investor	58,000,000 A-shares	580,000,000	50.4
Total	115,000,000 A-shares 1 B-share	1,150,000,000 1,000	100.0

[268] Shareholders Agreement, AS Tallinna Vesi, Draft 24 November 2000. [hereinafter, Shareholders Agreement].

[269] *See* Information Memorandum, *supra* note 259 at 7, A Shareholders Agreement (SHA) between the City of Tallinn and the Private Operator/Strategic Investor defining the roles and responsibilities of the parties, voting rights and board membership, dividend policy, the link between the SHA and the Services Agreement, Termination, and approvals of the City to any changes in the Company's shareholder structure, including Investor's share transfers. Investor's management control over the Company is subject to the City of Tallinn right of veto on a certain number of issues considered of fundamental interests to the citizens of the City.

[270] City of Tallinn and Tallinna Vesi. Services Agreement for (a) ensuring the functioning and maintenance of the public water supply and sewerage system of Tallinn, (b) extraction of fire fighting

Investor would conduct himself. The sale process would be defined under a *Share Sale and Subscription Agreement*.[271]

It must be noted that this transaction did not have a limited time-frame. In PPP arrangements like a Concession contract as in the case of Sofia, Bulgaria there is a time limit or duration of the contract, typically 30 years.

The obligations of the Private Operator/Strategic Investor are enshrined within the Shareholders Agreement and Service Agreement. These can be broadly defined as: having responsibility for the infrastructure and all assets of ASTV, operate, maintain and develop the Tallinn water supply and sewerage system, design, finance, construct and commission all works required for rehabilitation and construction of the water and wastewater system in accordance with National and EU standards and, in particular, with the WFD. There was a notable exception relating to storm water drainage systems which would be financed directly by the City of Tallinn with no recourse to the existing tariff.[272] Also there was a clear obligation to meet prescribed Levels of Service and other associated specified outputs and deadlines, and to establish performance indicators in water supply, wastewater collection and treatment, surface water drainage and other areas of ASTV's operations, to be achieved in the first 5 year term from 2001 to 2005.

In order to meet the specifically defined levels of service and outputs, the PO/SI was also required to finance investments estimated to be up to US$ 100 million.[273&274]

---

water from fire hydrants in the public water supply system and extraction of water from public water extraction points and (c) channelling of rain water, drainage water and other soil and surface water from public roads, streets and squares to the public sewerage system and the treatment thereof. Draft 24 November 2000, [hereinafter, Services Agreement].

[271] Tallinna Linn AS Tallinna Vesi and [Investor], Share Sale and Subscription Agreement, Final Draft 24 November 2000. [hereinafter, Share Sale Agreement].

[272] *Id.,* 67, Although the City of Tallinn should fund storm water drainage, it would be the responsibility of ASTV to use good business practices to implement the funding a manner that would enhances its ability to deliver services to the public and meet its operational obligations under the Service Agreement.

[273] As the author was one of the advisors to the CoT the following provides an explanation of the rationale behind the development of the CoT structure. It is important to note that in relation to the required investments, the intention of the CoT was to provide bidders with an indication of the amount of investment needed to meet the levels of service and performance indicators. However, the Service Agreement is not prescriptive about this amount; it is simply concerned with the outputs of the PO/SI. The way in which the PO/SI achieves these objectives is left to his discretion. Therefore, this PPP project is considered to be *output* and not *input* driven.

[274] Telephone interview with Mr Tim Lowe, Project Manager for the Tallinn, Estonia on behalf of International Water. (23 March 2003). Mr Lowe indicated that in his opinion, the key outcomes of

8.4 APPLICATION OF FRAMEWORK 1: EU COMPETITION LAW

Test 1 is to determine the *extent of public involvement.*

C1. Extent of public Involvement	• Role of the public and its involvement ➢ Has an exclusive right granted ➢ Extent of public control and compliance with WFD principles ➢ Structure of PPP chosen and rationale for choice • A PPP contract permitting equality and transparency to all bidders

The selected PPP structure was a *Partial Divestiture.* This meant that ownership was shared with the private sector. The divestiture meant that the CoT is a co-owner of the water utility. Therefore the test is: *'Is there public interest and is ownership by the public sector a controlling interest?'* The public's ability to exercise control was limited to the provisions of the Shareholders Agreement[275] and the Services Agreement[276] and limited to certain veto rights,[277] keeping in mind that the CoT was not a majority shareholder.[278] An exclusive right was granted to the Private Operator for an unlimited period of time. The Public Water and Sewerage Act[279] define the role of the public

---

this process can be summarised as a *'fast-track'* bidding procedure which was completed in a total of nine-months from initial notice to bid, no tariff increase in the first year and pre-defined tariff increases thereafter until year five, a commitment to meet WFD standards and a prescribed timetable in which to do so, retraining and outsourcing, improved customer service, investment continuity, flexibility of shareholding structure e.g. shares to employees, public offering and a pragmatic approach to asset renewal through first five year investment, mainly of above ground assets and extension of network (water and sewerage) to provide 100% coverage of services.

[275] *See* Shareholders Agreement, *supra* note 269.

[276] *See* Services Agreement, *supra* note 271.

[277] *See* Transfer of Shares, Shareholders Agreement, *supra* note 269. These 'veto' rights refer mainly to change in the Private Operator ownership.

[278] *See* Information Memorandum, *supra* note 259, at 5: CoT controlled 49% of the water utility.

[279] Public Water Supply and Sewerage Act, February 10th 1999, (entered into force March 5, 1999) [hereinafter the PWSSA]. Passed by the Riigikogu (Parliament) this Act governs rights and obligations of the state and local governments to supply consumers with water and treat their wastewater.

106   Public and Private Participation in the Water and Wastewater Sector

concerning development plans,[280] connections,[281] tariff setting[282] and relations between the Company and customers.[283]

The Shareholders Agreement necessitated careful negotiations as the CoT was not keen to lose control of the Company. It was felt by the Bidders that certain powers of the CoT were restrictive to the daily operations of the Company (ASTV) and the PO/SI sought more flexibility in this regard.

Although significant investment had been made by the company, in particular to water treatment facilities, it was apparent that significant investments were needed to meet the Water Framework Directive and its key objectives. Thus, it was decided that the PPP structure should allocate the risk of operation and achievement of specific objectives to the private sector through a transaction concerned with the definition of *outputs,*[284] i.e. a measure that would ensure compliance with EU legislation, in particular the WFD, and leave the *'how'* or *'the inputs'* to the PO/SI. This was considered to be an adequate balance of risk between the parties.[285]

The letting process was designed for transparency and fairness to all bidders. The CoT was under significant pressure to complete the transaction by the first quarter of 2001. Thus, it was imperative that the tender process be simple and risks clearly allocated to avoid lengthy negotiations. The tender process consisted of two stages, namely a pre-qualification[286] round, started in June 2000, followed

---

[280] *Id.,* Art 4 Section 1 PWSSA states that the Tallinn City Council shall approve a 12 year development plan for a public water supply and sewerage system.

[281] *Id.,* Art 6 Section 1 PWSSA states that the Tallinn City Council shall establish rules for connections to the public water and sewerage system.

[282] *Id.,* Art 6 Section 2 PWSSA states that the Tallinn City Council can determine a maximum rate for connections.

[283] *Id.,* Art 8 Section 4 PWSSA, Tallinn City Council has enacted regulations No 25, dated 25 August 1999 which approved the Rules on Use of Tallinn Public Water Supply and Sewerage System.

[284] *See* Obligations of the Company, Services Agreement, *supra* note 271.

[285] *See* Telephone interview with the Chairman of the Management Board and Director of Development, Mr Kalle Tiiter, *supra* note 263. Mr Tiiter commented that the fact that the transaction was successful and that there were a significant number of bidders at the pre-qualification stage, demonstrated that the 'architecture' of the PPP structure was sound and that the Private Sector was prepared to take on-board the risk allocation between the PO and the Public Sector.

[286] *See* Information Memorandum, *supra* note 259, at 9: Pre-Qualification Criteria consisted of three main criteria, namely *Criteria 1 – Water and Wastewater Experience,* whereby the International Private Operator had to demonstrate that it exercises management control in a company (or itself is a company) supplying water and sewerage services to at least 500,000 consumers, and that it exercises management control in water and/or wastewater companies in at least four (4) different countries. *Criteria 2 – Financial Capability* was demonstrated by an average annual turnover from provision of

by a four month round for final submission.[287] *Final submission* required the pre-qualified bidders to provide (a) an unconditional offer for the price of the 28,000,000 A-shares of ASTV to be sold and 30,000,000 A-shares to be subscribed and (b) a proposal for the yearly price change coefficients for the water and wastewater services within the Operating Region for the next five years.[288]

The award criterion was based on the best price for the sale and issue of the A-shares, together with the best coefficients for the yearly price change in the water and wastewater services for the next five years (60 per cent).

The second test is to determine whether there is an *Abuse of Dominant Position*.

C2. Abuse of Dominant Position	• Market share greater than 70–80% constitutes dominant position • Monitors and controls to protect consumers from abuse of dominant position. • Monitors and controls to protect the environment from abuse of dominant position • Mechanism to control changes in ownership or operation of private sector businesses involved.

The Company has a customer base of approximately 405,000 for drinking water. This coverage equates to approximately 99 per cent of the population. Wastewater and sewage collection systems cover approximately 385,000 consumers, which is approximately 95 per cent. This constitutes a dominant position.

Monitors and controls to protect consumers from abuse of dominant position: Pricing Controls were built into the tariff structure[289] specified in the

---

water services of at least US$17 million, and finally *Criteria 3 – Legal Documentation of the Company or grouping* showing their legal status.

[287] *See* Telephone interview with the Chairman of the Management Board and Director of Development, Mr Kalle Tiiter, *supra* note 263. Initially there were seven interested Parties which resulted in the following pre-qualified bidders: International Water, Lyonnaise des Eaux/Northumbrian, Vivendi/RWE Ruhrwasser.

[288] *See* Information Memorandum, *supra* note 259, at 68. In support of the coefficients of the yearly change in price for the water and wastewater services in the Operating Region for the next five years, Bidders were required to present a detailed business plan providing details of the manner in which the prescribed levels of service and performance indicators would be achieved, as well as financial projections of ASTV including financing necessary to achieve compliance with these indicators. In addition, the Bidder had to guarantee the completion of specific minimum underground e.g. network extension.

[289] *See* Public Water Supply and Sewerage Act, *supra* note 280. The basic premise for tariffs pricing control to include the introduction of a fixed fee element in the tariff as required by

Services Agreement.[290] The MMU also ensures competitive tariffs in comparison to benchmark sites elsewhere in Europe, as well as ensuring no discriminatory pricing among customers. Considerable emphasis was placed on technical strategies to meet the pre-defined levels of service criteria. Although the MMU covers functions to ensure suitable levels of customer service, it does not include customer advocacy functions.[291]

The winning bidder froze tariff increases for the first year and then continued with tariff increments slightly above inflation.[292] During the initial five year period additional tariff increases were not allowed except those related to inflation as stated in the Service Agreement. After the initial five year period the CoT and the PO/SI agreed to a tariff review to fix tariffs for another five year term and reviews thereafter every five years.[293]

Monitors and controls to protect the environment: Estonia had already established legislation to ensure the control of wastewater and pollution, with sanctions laid down.[294] The Tallinn Environmental Agency was responsible for monitoring environmental performance and to report to the city council. The quality of drinking water was monitored by the Tallinn Health Protection Authority[295] in an existing arrangement, as well as through provisions in the Services Agreement.[296]

Mechanism to control changes in ownership or operation of private sector businesses involved: The PPP structure envisaged mechanisms to control any changes in the ownership of the private operator to be governed through the Shareholders Agreement.[297] This contract not only details the relationship between the Private Operator and the CoT but also details certain issues such as rights and responsibilities,[298] dividend policy, changes in ownership etc.

Section 14.1.1, setting a non discriminatory price for services for domestic and non-domestic clients in line with Section 14.4 and setting tariffs allowing for inflation increases.

[290] *See* Tariffs, Services Agreement, *supra* note 271.

[291] *See* Review Procedures & Service Level Test and Auditing, Services Agreement, *supra* note 271. This means that the MMU can only act against the Private Operator/Strategic Investor if there is a direct breach of the levels of service and outputs prescribed in the Service Agreement. The MMU cannot act on behalf of customers in disputes that may occur between the service provider and an individual customer.

[292] Although the tariff increments were higher than inflation, the exact values could not be revealed due to confidentiality reasons. When International Water and the City of Tallinn were approached for an indication as to these levels, they declined to comment.

[293] *See* Tariffs, Services Agreement, *supra* note 271.

[294] *See* Article 3, Water Act, *supra* note 296.

[295] Water Act, May 11 1994, Riigikogu Proceedings, TNN 1564 (1994), [hereinafter, Water Act].

[296] *See* Service Level Test and Auditing, Services Agreement, *supra* note 271.

[297] *See* Article 4: Transfer of Shares, Shareholders Agreement, *supra* note 269.

[298] *Id.*, Art 3.

Tallinn, Estonia partial divestiture/sale contract

In particular, the Shareholder Agreement included a clause whereby the City of Tallinn required its approval for any changes in the shareholder structure of the Company, including investor's share transfers. This included a five year moratorium before shares could be sold, together with clauses related to the service levels and criteria required of any potential new operator.

The third test was to determine **Relevant Market.**

C3. Relevant Market	• Geographical market defined • Significance of activities within this defined market • Product market – any interchangeable substitutes

The Services Agreement defines the market area.[299] However, the Company's activities are significant in this context, as the Company was the only operator of the water and wastewater sector in the City of Tallinn. As the product market is water and wastewater services, there are no interchangeable substitutes although there are existing wells on some properties.

The fourth test relates to **State Aid.**

C4. State Aid	• Any state aid that distorts competition when bidding
	• Exceptions:
	  ➤ Aid for social development
	  ➤ Aid for economic development
	  ➤ Promoting major European interest

At the time of the bidding/tender process, the Government of Estonia and the City of Tallinn did not influence the competition nor in any way was it influenced. Award criteria were based on best price for the A-shares to be sold and issued as well as best coefficients for tariff increases. These were opened in full public view.

State Aid involvement was included in three ways: (a) aid for social development was provided by a CoT subsidy of up to a maximum of 80% for the total water and sewerage connection fees for the Company's customers;[300] (b) aid for social development was also granted through the CoT funding the extension of the storm water drainage system (also known as surface water drainage) which was not considered to be the responsibility of the Company and for which the

---

[299] *See* Mandate, Services Agreement, *supra* note 271.

[300] *See* City of Tallinn Resolution, *supra* note 264. This resolution was also supported by the City of Tallinn Decree No. 25 dated 15 June 2000, in which a certain amount of the proceeds from the sale of the existing shares of the Company were to be used for financing, sewerage and water network connections, surface water drainage network extension and necessary road excavation works for the network extensions.

110   Public and Private Participation in the Water and Wastewater Sector

Company received payment; and (c) aid was granted for economic development because of the timing required by the Company to achieve full compliance with the EU standards for wastewater treatment.

The fifth test is to determine if there is a *regulatory mechanism to support the Competition principles*.

C5. Regulatory mechanism to support Competition principles	• Mechanism to prevent other undertakings from entering market • Mechanism to accommodate changes in Community law or *acquis communautaire*. • Mechanism to control changes in national government policy within European law. • Mechanism to ensure eventual competition through limited duration contracts, or mechanisms encouraging competitors to enter the market at the same level as incumbent operator.

The contract does not specifically prevent other undertakings or firms from entering the designated area. The chosen PPP structure envisages a relationship between the parties governed by a contractual mechanism defined in a Service Agreement which will detail obligations in operation, maintenance, billing and collection, capital investment and other issues pertaining to the running of the water and wastewater services for the 'designated area'.

The Contract Annexes are very specific as to the terms of service delivery and obligations expected of the Concessionaire. These include obligations to meet quality criteria for both water and wastewater which, at the time of the signing of this agreement were in line with accession criteria. However, the contract does allow for a 'Change in law' provision that seeks to address any changes that may occur due to Community law.[301]

Is there a mechanism to control changes in national government policy within European law? Under European law a contract cannot limit a sovereign government's ability to issue new laws, regulations and decrees. Thus the PO is

---

[301] Telephone interview with Mr Priit Koit, Director of Suprema Securities, Merchant banker responsible for the sale process of AS Tallinna Vesi (1 Mar 2000). Mr Koit stated that there are a number of clauses in the contract between the City of Tallinn and the operator, which require the operator to meet all changes in law as part of his obligations. These state that the operator must meet these changes at his cost, unless a change during the operating year, results in a change in operating cost more than 5% over one operating year, or 7.5% over two years. Should this be the case, there is provision to review the tariffs and funding requirements so that the operator can meet targets profitably.

obliged to implement any changes in law. However, whilst this right of Government is not challenged, the contract does specify a mechanism to allow for changes in law[302] that may have an adverse financial effect on the PO i.e. if, as a result of the implementation of that change in law, there is a significant variation in the PO's financial position. In this case the PO has a right to seek recompense through an increase in tariff or any other mechanism to be decided by the CoT.

The nature of the selected PPP meant that the private operator is the biggest operator in the area and as such, there is no direct competition. The lack of competition is mitigated by tariff levels negotiated on an ongoing basis, based on benchmark sites elsewhere in Europe. It should be noted that the CoT does not prevent the PO from expanding its service area beyond the administrative area covered by the Services Agreement.[303]

8.5 APPLICATION OF FRAMEWORK 2: REGULATORY PROVISIONS

The second framework for analysis relates to regulatory provisions. In this context the first of the test is to determine if the PPP contract is *well defined and is predictable*.

R1. Predictability/well defined PPP contract	• Structure used to assure sustainability • Current contractual structures • Adequate allocation of roles, rights, responsibilities and obligations between parties.

Initially *"there was an unacceptable level of regulatory uncertainty upon which a Bidder's financial projections could be made"*.[304] Thus it was necessary to create a mechanism that allowed regulatory certainty for investors. This was achieved through the creation the Mandate Monitoring Unit (MMU).[305]

The regulatory structures in place are considered in the Service Contract where the MMU's functions are broadly defined[306] to assist in the CoT

---

[302] *See* Changes in Law, Services Agreement, *supra* note 271.
[303] *See* Obligations of the City, Art 7, Services Agreement, *supra* note 271.
[304] *See* Telephone interview with Mr Priit Koit, *supra* note 302.
[305] *See* Obligations of the City, Art 3, Services Agreement, *supra* note 271.
[306] *Id.*, Art 4 and Art 5.

compliance monitoring of the fulfilment of the PO/SI's obligations as stipulated in the Service Agreement. Roles and responsibilities were adequately defined in the Services Agreement.[307] However, there was a need to clarify the tariff setting procedure. It was resolved that, as the MMU was not legal party to the Service Agreement, it could therefore not be granted the right to set tariffs. Thus the MMU's role was to provide the CoT with recommendations in the tariff setting procedure.[308]

The second test relates to the *alignment of the PPP contract with National law.*

R2. Adequate alignment of the PPP contract with national law	• Establish broad principles of regulatory policy using specific laws of the general and national water-sector • Law to set national service standards • PPP contract complimentary to such national law and covering specific commercial details of area served • Establish predictability and enforcement of the PPP contract, i.e. safeguards incorporated into the contract particularly relating to protecting public interest. • Any national legislation enabling alignment of accession criteria

The legal framework that underpins involvement of the private sector is the Public Water Supply and Sewerage Act[309] and the Water Act[310] apart from specific City Council resolutions permitting private sector participation in the water and wastewater company.[311]

---

[307] *Id.,* Section 4: Obligations of the Company and Section 5: Obligations of the City.

[308] *See* Telephone interview with Mr Priit Koit, *supra* note 302. Mr Koit said that one of the reasons why the tariffs were set for the first five years is that because there was no independent Estonian regulator, the City Government would act as the water tariff regulator for the first five years, beginning from 2001. However, since the City is one of the Parties to the Services Agreement, and is intimately involved with the proposed transaction, the City Government could not properly be regarded as an independent and impartial regulator.

[309] *See* Water Act, *supra* note 296.

[310] *See* Public Water Supply and Sewerage Act, *supra* note 280.

[311] *See* City of Tallinn Resolution, *supra* note 264.

National law already included quantitative and qualitative standards in line with the WFD and the Helsinki Convention on Baltic Sea Environmental Protection.

The Service Agreement is in line with national law (specifically the Water Act) and covers specific commercial details in the area served.

Predictability and enforcement of the regulatory functions is established in the Service Agreement, where fines and penalties to be applied in the event of non-compliance are very specific.[312]

National enabling legislation had been adopted prior to the introduction of the private sector investor.

The third test is the contract's *flexibility/adaptability to deal with long contract periods.*

R3. Degree of flexibility/ adaptability to deal with long-term contracts	<ul><li>Enable flexibility for adaptation and evolution for reaction in changing circumstances,</li><li>Ensure no attempt of unfair advantage taken by any party should changes be required in the contract</li><li>Impartial supervision to ensure the original duties and obligations</li></ul>

In this context, flexibility has been achieved through the introduction of the MMU as a means of enabling regulation and dealing with changes during the life of the contract (see the discussion below). It must be noted, however, that the value of the shares at Termination required careful consideration, and a mutually acceptable formula had to be devised.[313]

Perceived unfair advantages in the contract are covered in a precise mechanism for *'exceptional items'*[314] in the Service Agreement for tariff reviews. This specifies when changes may be made and the procedure to apply them.

---

[312] *See* Information Memorandum, *supra* note 259, Penalties, 86.

[313] Unfortunately due to the sensitive nature of this termination value of shares mechanism, the author is prevented from commenting on this aspect of the transaction.

[314] *See* Services Agreement, *supra* note 271. *'Exceptional items'* in the Service Agreement were considered those items that were not covered under a 'Normal' Tariff review procedure and could include issues such as inflation escalation beyond agreed levels, extensions to the network beyond those agreed in the Development Plans etc.

114   Public and Private Participation in the Water and Wastewater Sector

Impartiality to ensure original obligations are met is achieved through the MMU's precise mandate, roles, responsibilities and obligations to which both public and private sectors subscribe and abide by.

The fourth test is whether or not *independence, accountability and transparency have been achieved.*

R4. Independence, accountability and transparency of regulation	• Necessity of an independent and fully competent regulatory body • Regulator to uphold public and private interests • Ensure eventual competition • Ensures public participation

The intention of the Government is to establish the principles of a national regulator. At the same time the MMU's role has been extended since the transaction took place in early 2001. It will monitor the PO/SI compliance in the City of Tallinn, but will also apply to other areas of service provision throughout the country. The main issue of contention was tariff setting. It was resolved that, as the MMU was not a legal party to the Service Agreement, it therefore could not be granted the right to set tariffs and its role was to provide the CoT with recommendations in the tariff setting procedure.

Although other water privatisations have been envisaged in Estonia, there has been no competition up to now. Should this occur, benchmarking could be established to introduce comparison of competitive key indicators.

Public participation was ensured at the time of bidding through consultative sessions. In the future, public participation will be ensured through the MMU's customer advocacy role that will eventually be transferred to the national regulator.

8.6 APPLICATION OF FRAMEWORK 3: WATER FRAMEWORK DIRECTIVE

As in the case of Bulgaria, the first test for the WFD is related to achieving water of good quality. The WFD is specific about what it means by "good status" and in this test we assess whether there are measure s in place to achieve this status (defined as 'Good Surface Waters' within Ecological & Chemical Criteria and 'Good Groundwater's' within Quantitative & Chemical Criteria) by 2015. In addition we seek to determine whether there are quantitative and qualitative criteria in place as well as to measure 'sustainable use'.

WFD 1. Water Quality – "Good Status"	Prevent deterioration of 'status' in all waters Achieve 'good status' (sustainable) in all waters by 2015 in: • 'Good Surface Waters' within Ecological & Chemical Criteria • 'Good Groundwater' within Quantitative & Chemical Criteria • Application of quantitative and qualitative criteria • Application of detailed criteria for measuring 'sustainable use' Extensions: – when necessary improvements 'cannot reasonably be achieved' Exceptions: Art4(5) non feasible or disproportionately expensive Art 4(6) temporary deterioration due to unforeseen natural circumstances Art 4(7) results in modifications of waters and benefiting human health, safety or sustainable development.

The Estonian government had already aligned its water legislation with EU legislation. A good indication of this was their inclusion in the Helsinki Convention on Baltic Sea Environmental Protection (HELCOM).

The contract needed to ensure that performance was monitored and non-compliance penalties were applied according to legislation. Thus the MMU was formed and its functions defined as above.

The second test related to *Management by river basin.*

WFD 2. Management by River Basin	River Basin Management Plans (Art 13) to include: • Characteristics of River Basin • Summary of significant pressures and human activity • Programme of measures • International River Basins Within EU – must produce a plan for Member State and co-ordinate with others in an overall plan Extending beyond EU – should endeavour to produce single plan, and at least plan for Member State

At first sight, it could be argued that the principle of management by river basin is not applicable in the case of Tallinn. Nevertheless, upon closer examination, it does have applicability in respect to developing and maintaining control measures in, e.g. water abstraction, pollution control etc. In addition, the contract required the private operator to report on any *Special characteristics* as defined in the WFD as well as an obligation to report any discharge into the sea. The third test relates to ***public participation.***

WFD 3. Public participation	Must ensure public participation (Art 14) and consultation, and be available for comments: • Plan timetable and work programme (3 yrs prior to adoption) • Interim overview of significant water management issues • Draft copies of River Basin Management Plan Public access to information and background documents

This was ensured by the involvement of City Councillors, each representing different parts of the City of Tallinn. Regular press conferences were held by the CoT at the time of bidding to inform the public. In addition, once the successful Bidder was identified, a press conference was held whereby full details of the tariffs and the performance indicators were presented for the following five years. Furthermore, the MMU had a clear mandate to protect the public interest through participation.[315] One of the obligations enshrined in the Services Agreement[316] was for the operator to support the relevant authorities, as appropriate, to develop river basin plans.

Fourthly, the test to determine whether there is an adequate ***monitoring and reporting programme*** is needed.

WFD 4. Monitoring and Reporting	Monitoring (Art 8) Must establish monitoring programme (2006) Include: • Surface water: volume and level, or rate of flow; ecological and chemical status; and ecological potential • Ground water: chemical and quantitative status • Technical specifications and standardised methods Reporting: (Art 15) Summary of reports of monitoring programmes implemented (Art 8) Review of impact of human activity on status of waters (Art 5) Economic analysis of water use (Art 5)

---

[315] *See* Obligations of the City, Art 4, Services Agreement, *supra* note 271.
[316] *See* Schedule H, Services Agreement, *supra* note 271.

The City of Tallinn had already established agencies to monitor drinking water quality and environmental issues.[317] Further controls were needed to protect service levels and tariffs to consumers. The MMU was established to ensure compliance with the relevant performance indicators and Levels of Service.[318]

Finally the fifth test related to *Cost recovery*.

WFD 5. Cost recovery	Take into account the principle of full cost recovery Follow an economic analysis Long-term forecast of supply and demand Volume, price and costs of water services Relevant investments In accordance with the 'polluter pays' principle Water pricing policies must provide adequate incentives for efficient water use by 2010 Must be adequate contribution of different water users including industry, agriculture and households to recover costs of water services.

Under public ownership, the water utility reported a profit and had been managed in line with private sector principles. This meant that the principles of full cost recovery were already being applied.[319] Moreover, under Estonian Law, the Company may apply penalties under the 'Polluter pays' principle. Water pricing policies are specifically included in the Public Water Supply and Sewerage Act.[320] Water pricing creates adequate incentives for efficient water management since the tariff should cover production costs, allow compliance with quality, safety and environmental protection standards, and provide for operating with a justified profit. The setting of the tariffs for the period after the bid will be achieved according to economic principles.[321] It must be noted that

---

[317] *See* Schedule H, Auditing Requirements, Services Agreement, *supra* note 271.

[318] *See* Information Memorandum, *supra* note 259, Section 5. Performance indicators were set at the time of divestiture. These directly relate to national law and standards. These in turn, reflect the WFD and other relevant EU Standards, including Helcom for discharges in to the Baltic Sea. Progressive performance standards were also set and thus established a compliance timetable.

[319] *See* Telephone interview with Mr Chris Shugart, *supra* note 262. Mr Shugart mentioned that AS Tallinna Vesi had received significant funding from EBRD. These were 'preferential rates' commercial loans whose covenants were very specific about the principle of full cost recovery through tariff and the avoidance of any subsidy or cross-subsidy.

[320] *See* Public Water Supply and Sewerage Act, *supra* note 280.

[321] *See* Information Memorandum, *supra* note 259. The tariffs for the first five years will be in accordance with the tender from the successful bidder. From then onwards tariffs will be set in a manner that promotes the economy and efficiency in carrying out the duties of the Company and that protects customer interests.

118   Public and Private Participation in the Water and Wastewater Sector

the tariff setting mechanism i.e. pricing structure, does not permit discriminatory pricing of domestic and non-domestic clients.

8.7 ESTONIA TALLINN SUMMARY OF FINDINGS

The multi-criteria framework (EU Competition Law, Regulation and WFD) was systematically applied to the Tallinn PPP structure. It must be noted that because this was a partial share sale, the overall contractual structure is slightly more complex than in a traditional concession agreement as was the case in Sofia, Bulgaria. Below we summarise how each of the issues of the framework have been dealt with by the Tallinn divestiture structure and the corresponding contractual structures of Shareholders Agreement and Operating Agreement:

1. **EU Competition Law** principles:

Extent of public involvement was fully achieved as the CoT was a shareholder of the new formed company ASTV. The relationship between the private and public sector was regulated through the Shareholders Agreement.

Equal treatment of bidders was achieved and all bidders were granted full access to information and facilities. The bidding was awarded on the basis of share price offered and proposed tariff for the next five years.

Given the monopoly rights that have been granted to ASTV, the assessment for Dominant Position demonstrates that adequate *protection for customers* has been put in place, through the obligations in terms of levels of service to customers. However it can be argued that, as the CoT is also a shareholder, there is an implicit conflict of interest between the CoT as a shareholder (looking to maximise its returns as a shareholder), and the CoT as an advocate for the customer. In addition *ownership* control mechanisms were established through the golden share rights granted to the CoT in the Shareholders Agreement.

Environmental protection was achieved as the company was already complying with the HELCOM standards (which are consistent with WFD standards) prior to the introduction of the private sector.

The test for relevant market requires an understanding of whether the company's activities are significant in a defined market area. Clearly, as ASTV is the only operator of the water and wastewater sector in the City of Tallinn, there are no interchangeable substitutes for a customer to obtain the same service.

State aid did exist in the for storm water drainage.

2. With regard to **Regulatory principles**, the divestiture structure allowed:

Partial *independence and accountability* through the establishment of a regulatory function (by contract) in the form of a body called the Mandate

Monitoring Unit (MMU). The issue of conflict of interest of the CoT as a shareholder is however an issue of concern.

However, *competition promotion* was not achieved directly through the contract but was mitigated in part through a competitive share sale and tariff structure (as award criteria) and later on through the use of the MMU.

Flexibility in regulatory functions is permitted in time, as the functions of the MMU are envisaged to be transferred to a national regulator once the necessary institutional structures are put in place.

3. **WFD requirements:**

With respect to *water quality* these have been fulfilled totally due to significant investments in ozonation technology. As regards wastewater, full compliance with WFD is envisaged.

The requirement for *management by river basin* is not expressly required as part of the Operating Agreement. However, the principles of water abstraction, pollution control and prevention, etc are implicit in the obligations enshrined in the Operating Agreement.

Monitoring and reporting functions are well established in the Operating Agreement, as are the principles of *cost recovery* which form the basis of the tariff regime being applied. In addition, the company was being run on full cost recovery principles before the divestiture.

It can thus be concluded that the form of PPP structure that has been chosen i.e. a divestiture, is an effective means to achieve compliance with the three categories analysed. The conditions preceding the Tallinn divestiture are unique and make their applicability to other Accession Countries less probably. Amongst these exceptional characteristics include a significant capital investment program undertaken prior to the introduction of the private sector (this did not mean that all capital investment needs were satisfied, particularly with regard to storm drainage), water treatment facilities using high-technology such as ozonation allowing water quality standards to be met immediately, and a fully operational customer cadastre.

As in the case of Sofia, Bulgaria, the full success of the divestiture and effectiveness cannot, at this stage, be determined.

9
Conclusions

This book has sought to establish that Private Sector Participation or PPP in the water and wastewater sector could play a significant role in reconciling the need of Accession Countries for development, funding and meeting the accession criteria in particular those of the Water Framework Directive. Furthermore, it delves into whether the mechanisms available to Accession Countries to 'partner' with the private sector provide a sound legal basis for future development. We have used two landmark case studies whose progress to meet pre-defined criteria, through the application of a multi-criteria framework, has been assessed and in so doing also meets EU accession criteria. In the conclusions drawn, the book implicitly explores the possibility of using similar structures in other Accession Countries.

Consideration has been given to the growth of the EU and the subsequent accession criteria that evolved, showing the importance of the water and wastewater sector in meeting the EU's environmental accession criteria. This is

© 2009 IWA Publishing. *Public and Private Participation in the Water and Wastewater Sector: Developing Sustainable Legal Mechanisms*, by X Cledan Mandri-Perrott. ISBN: 9781789065299. Published by IWA Publishing, London, UK.

reflected in the Water Framework Directive which enshrines the basic principles for management and the sustainability of the sector both within and outside the EU.

The book explores the various PPP options available to Governments. Depending on the real/actual participation of the public sector in the PPP form selected, it highlights the need to determine whether the public sector continues to exercise a level of control and economic interest in the undertaking. Once this has been determined, it continues with an assessment of, the key EU Competition principles of Abuse, Dominant Position, Relevant Market and Affecting Trade between Members States in which it concludes that EU Competition law plays an important role in shaping PPP structures. Moreover, it reveals that a sustainable arrangement depends on the balance between the contractual framework which defines the form of PPP and the national and Community law that complements the public/private partnership.

Similarly, in many cases, the provision of water and wastewater services results in a natural monopoly,[322] and so there is a need for adequate regulation to ensure the sustainable development of the sector, and an arrangement which complies with Community law and the WFD and is fair to the customer, the private party and the public sector.

Therefore, we examined and explored in detail the EU Competition law, the Water Framework Directive and the principles of Regulation to provide the basis for the multi-criteria qualitative framework for analysis. This analytical framework was applied to two landmark case studies in Eastern Europe, both of which were Accession Country candidates at the time this book started. Since then, Estonia has gained accession and Bulgaria joined in 2007. The case studies are Sofia-Bulgaria and Tallinn-Estonia respectively. Each case study was selected because of their differing characteristics such as population, geographical positions and the applicability of WFD,[323] as well as their position with respect to meeting EU accession criteria. Estonia was far more advanced in this process than was Bulgaria, but both countries and their respective capital cities, which are the focus of this book, shared the same obligation to comply with the Community law principles and the WFD.

The multi-criteria framework (EU Competition Law, Regulation and WFD) was systematically applied to each country's chosen form of PPP, to assess the

---

[322] Usually there is no more than one provider of water and water services in a given region, therefore giving rise to a 'monopoly situation'. However, there are cities in which there are two distinct service providers. In either event the provision of water services is not open to economic conditions and as such, there is a risk of 'monopoly'.

[323] The City of Sofia, Bulgaria is land locked, and the City of Tallinn, Estonia, discharges directly into the Baltic Sea.

Conclusions 123

effectiveness with which each country deals with accession criteria (WFD), the legality of the chosen structure and its sustainability through effective regulation. These results are summarised in the table below.

RESULTS OF ANALYTICAL FRAMEWORK FOR PPP CASE STUDIES

	Bulgaria	Estonia
EU Competition law requirements fulfilled		
Extent of public involvement	Not directly	Yes, part owner
Equal treatment of bidders	Yes	Yes
Dominant Position		
Protection of customers	Yes	Partly[324]
Protection of Environment	Partly[325]	Yes[326]
Ownership control	Yes[327]	Yes[328]
Relevant market[329] – are there any interchangeable substitutes	No	No
State Aid	Yes	Yes
Regulation principles fulfilled		
Independent, accountable etc	Partly	Partly
Alignment of PPP with National law[330]	Yes	Yes
Promoting competition[331]	Not directly	Not directly
Eventual Flexibility (in time)	Yes	Yes

(*continued*)

---

[324] The protection of customers is achieved through setting pre-determined service levels which include attention to customers.

[325] The Concession Agreement allowed for specific 'waivers' to environmental obligations recognising the particularly bad current state regarding wastewater treatment and discharge. In addition, EBRD provided funding for wastewater treatment works.

[326] Prior to the introduction of PPP in Estonia, the water utility was already obliged to comply with the Helsinki Convention on Baltic Sea Environmental Protection HELCOM.

[327] Both contracts allowed for changes in ownership, as well as changes in the Private Operator subject to a '*suitably qualified substitute*'.

[328] There was a five year moratorium period in which Private Operator ownership changes were not permitted but dilution of the City's interest was.

[329] Relevant market test is: given a defined market, what is the significance of activities? For both contracts the activities are significant as they constitute a dominant position in the defined market.

[330] This includes changes in national law as well as changes in Community law including WFD.

[331] The lack of competition was mitigated by fixed tariff levels for the initial five years and subsequent-ongoing negotiations with the City of Tallinn based on benchmark values in similar countries in Europe.

	Bulgaria	Estonia
WFD requirements fulfilled		
Water quality	Partly	Yes
River Basin	Partly[332]	No[333]
Monitoring & reporting	Yes	Yes
Cost recovery	Yes	Yes[334]

The PPP forms chosen prove, in principle, to be adequate for the context of each city. For example, Sofia, Bulgaria could not have realistically been able to entertain a divestiture as the private sector was not interested in committing to both a capital investment programme with full operational risk, as well as purchasing shares. At the same time a service or management contract would not satisfy government's inability to raise funds to meet WFD standards, nor were the funds available to enter into a lease type contract. A BOT arrangement (and any of its variants) could have resolved a specific problem, for example sewage treatment for the City of Sofia, but failed to offer a comprehensive and integrated strategy for both water and wastewater services. Thus the concession structure was chosen.

Similarly, the City of Tallinn was in the privileged position of being able to offer shares in a company that was operating at a profit (albeit small), having already spent considerable monies to deal with compliance of both water and wastewater standards. Investor interest in a share sale was high and a concession structure (nor indeed any other form of PPP) would not have fully capitalised on the significant benefits that the public company had achieved to date, or been able to mobilise the significant levels of capital investment required to ensure full compliance with WFD standards.

Given that there was no national regulator in place prior to the introduction of the PPP structure, the book placed significant emphasis on the regulatory aspects of the provision of water services, and as such, examined the regulatory conditions for the provision of water services in these two countries. It was

---

[332] Some aspects have been put in place e.g. requiring the Concessionaire to identify *'special characteristics'*. However, there are still more actions required from Government to establish the administrative arrangements within each river basin.

[333] On first sight, it could be argued that this principle is not applicable to Tallinn. However it is applicable when referring to ensuring control measures in, e.g. water abstraction, pollution control etc. *Special characteristics* should be reported, together with any discharges into the sea.

[334] It must be noted that the water utility in Tallinn, Estonia, unlike Sofia, Bulgaria, had already been managed under cost efficient and thus cost recovery, principles.

found that although not all conditions were met, nevertheless, they provide adequate regulatory controls through their respective contractual structures.

It is still too early to assess whether PPP has been an effective tool for WFD compliance for both Estonia and Bulgaria. Additionally, it is difficult to gauge to what degree PPP has been the only factor in helping these two countries achieve their accession ambitions. Certainly Estonia was more advanced in meeting the WFD than Bulgaria – even prior to the introduction of the private sector. For Bulgaria, having reached accession in 2007, the advancements in terms of quality and quantity of service provision in Sofia have arguably been a contributor to achieving accession. For both cases, the jury is still out. The WFD, by its very nature and overall definition, cannot be accomplished immediately. This, in fact, could be an area for further research in say five years. Nevertheless, the application of the multi-criteria framework for analysis that is presented in this book does suggest that the chosen forms of PPP go a significant way to meet basic principles of Community law and have the potential, therefore, to help both countries meet WFD requirements.

The general conclusions drawn from these two examples are that they demonstrate that any PPP development must to go hand in hand with a convergence of private, national and trans-national objectives, together with the resulting regulatory framework, to achieve accession. It further concludes that the structure of PPP chosen requires careful consideration and that the PPP contracts should be flexible, creative and manageable. It demonstrates that these can be sustainable through effective contractual, legislative and regulatory mechanisms which will ensure an adequate balance of risk between the public and private sectors.

While PPP is therefore not the only route towards EU accession, it does permit financial, managerial, technical and commercial expertise to be brought to bear. Accordingly, there are a number of general lessons that can be learnt from this book and applied to other Accession Countries:

Firstly, there are no right or wrong PPP structures. The structure chosen must reflect, as best possible, the actual realities on the ground and the conditions of the water utility. The key advantage here is that the finish line is known – i.e. compliance with WFD. The manner in which this is achieved can vary depending on the decision to involve the private sector or not.

Secondly, whatever form of PPP is chosen, and since there is a natural tendency for a monopoly position in a water utility, Community law principles, particularly competition law, should be considered.

Finally, for PPP to be sustainable, the strategy chosen must converge with private, national and trans-national objectives and be coupled with adequate national and international regulatory frameworks that ensure compliance, transparency and fairness.

AREAS FOR FUTURE RESEARCH

As mentioned above, it is too early to generalize about the effectiveness of PPP structures. The framework used in this book can be expanded to help design structures and explore their possibilities and limitations in the drive towards compliance with the WFD and eventual accession to the EU.

It is outside the scope of this book to analyse the ways in which the limitations can be addressed. For example; How could the public sector restructure to create the same advantages as one of the PPP structures examined? How to balance the profit motivation of the private sector with government's obligation to provide services to the poor? How are subsidies managed?

This book does not explore how other EU Directives might affect the provision of water services. Similarly, the book does not delve deeply into the regulatory aspects of setting tariffs. It would be useful to apply the same multi-criteria framework for analysis to these two cities in, say, five years to determine what real impact the PPP structures have had in meeting WFD requirements.

Overall, I hope that this book has provided you with an insight as to what options are available for PPP and how these can be made consistent with EU law. The EU has taken some significant strides towards recognising PPPs as a valid tool to bring to bear operational, technical, financial and other resources. Combining the private and public sectors can be a useful tool. However, as said before, PPP is no panacea. Good public utility management of water and wastewater can be as effective as the private sector. Key for success is to create an enabling environment where the rules, roles and responsibilities of all parties, including all stake holders, are clearly defined. Good luck!

THE END

Annex 1

FRAMEWORK 1: EU COMPETITION LAW

Competition Principle	Indicator for inclusion in analytical framework	Competition Principle	Indicator for inclusion in analytical framework

C1. Extent of public Involvement	• Role of the public and its involvement
	   ▵ Has an exclusive right granted
	   ▵ Extent of public control and compliance with WFD principles
	   ▵ Structure of PPP chosen and rationale for choice
	• A PPP contract permitting equality and transparency to all bidders
		C4. State Aid	• Any state aid that distorts competition when bidding
			• Exceptions:
			   ▵ Aid for social development
			   ▵ Aid for economic development
			   ▵ Promoting major European interest

C2. Abuse of Dominant Position	• Market share greater than 70–80% constitutes dominant position • Monitors and controls to protect consumers from abuse of dominant position. • Monitors and controls to protect the environment from abuse of dominant position • Mechanism to control changes in ownership or operation of private sector businesses involved.	C5. Regulatory mechanism to support Competition principles	• Mechanism to prevent other undertakings from entering market • Mechanism to accommodate changes in Community law or *acquis communautaire*. • Mechanism to control changes in national government policy within European law. • Mechanism to ensure eventual competition through limited duration contracts, or mechanisms encouraging competitors to enter the market at the same level as incumbent operator.
C3. Relevant Market	• Geographical market defined • Significance of activities within this defined market • Product market – any interchangeable substitutes		

Annex 2

FRAMEWORK 2: REGULATORY FRAMEWORK FOR ANALYSIS

Regulatory Principle	Indicator for inclusion in analytical framework	Regulatory Principle	Indicator for inclusion in analytical framework
R1. Predictability/ well defined PPP contract	• Structure used to assure sustainability • Current contractual structures • Adequate allocation of roles, rights, responsibilities and obligations between parties.	**R3. Degree of flexibility/ adaptability to deal with long-term contracts**	• Enable flexibility for adaptation and evolution for reaction in changing circumstances • Ensure no attempt of unfair advantage taken by any party should changes be required in the contract • Impartial supervision to ensure the original duties and obligations
R2. Adequate alignment of the PPP contract with national law	• Establish broad principles of regulatory policy using specific laws of the general and national water-sector • Law to set national service standards • PPP contract complimentary to such national law and covering specific commercial details of area served • Establish predictability and enforcement of the PPP contract, i.e. safeguards incorporated into the contract particularly relating to protecting public interest. • Any national legislation enabling alignment of accession criteria	**R4. Independence, accountability and transparency of regulation**	• Necessity of an independent and fully competent regulatory body • Regulator to uphold public and private interests • Ensure eventual competition • Ensures public participation

Annex 3

FRAMEWORK 3: WATER FRAMEWORK DIRECTIVE (WFD)

WFD Principle	Indicator for inclusion in analytical framework	WFD Principle	Indicator for inclusion in analytical framework
WFD 1. Water Quality – "Good Status"	Prevent deterioration of 'status' in all waters Achieve 'good status' (sustainable) in all waters by 2015 in: • 'Good Surface Waters' within Ecological & Chemical Criteria • 'Good Groundwater' within Quantitative & Chemical Criteria • Application of quantitative and qualitative criteria • Application of detailed criteria for measuring 'sustainable use' Extensions: – when necessary improvements 'cannot reasonably be achieved' Exceptions: Art 4(5) non feasible or disproportionately expensive Art 4(6) temporary deterioration due to unforeseen natural circumstances Art 4(7) results in modifications of waters and benefiting human health, safety or sustainable development.	WFD 2. Management by River Basin	River Basin Management Plans (Art 13) to include: • Characteristics of River Basin • Summary of significant pressures and human activity • Programme of measures • International River Basins Within EU – must produce a plan for Member State and co-ordinate with others in an overall plan Extending beyond EU – should endeavour to produce single plan, and at least plan for Member State

Annex 3

WFD 3. Public participation	Must ensure public participation (Art 14) and consultation, and be available for comments: • Plan timetable and work programme (3 yrs prior to adoption) • Interim overview of significant water management issues • Draft copies of River Basin Management Plan Public access to information and background documents	WFD 5. Cost recovery	Take into account the principle of full cost recovery Follow an economic analysis Long-term forecast of supply and demand Volume, price and costs of water services Relevant investments In accordance with the 'polluter pays' principle Water pricing policies must provide adequate incentives for efficient water use by 2010 Must be adequate contribution of different water users including industry, agriculture and households to recover costs of water services.
WFD 4. Monitoring and Reporting	Monitoring (Art 8) Must establish monitoring programme (2006) Include: • Surface water: volume and level, or rate of flow; ecological and chemical status; and ecological potential • Ground water: chemical and quantitative status • Technical specifications and standardised methods Reporting: (Art 15) Summary of reports of monitoring programmes implemented (Art 8) Review of impact of human activity on status of waters (Art 5) Economic analysis of water use (Art 5)		

Annex 4

© 2009 IWA Publishing. *Public and Private Participation in the Water and Wastewater Sector: Developing Sustainable Legal Mechanisms*, by X Cledan Mandri-Perrott. ISBN: 9781789065299. Published by IWA Publishing, London, UK.

136   Public and Private Participation in the Water and Wastewater Sector

DIAGRAMMATIC REPRESENTATION OF TEST

The following diagram represents the two possible routes that result from applying the test to different forms of PPP structures, i.e. one a concession structure and the other a divestiture.

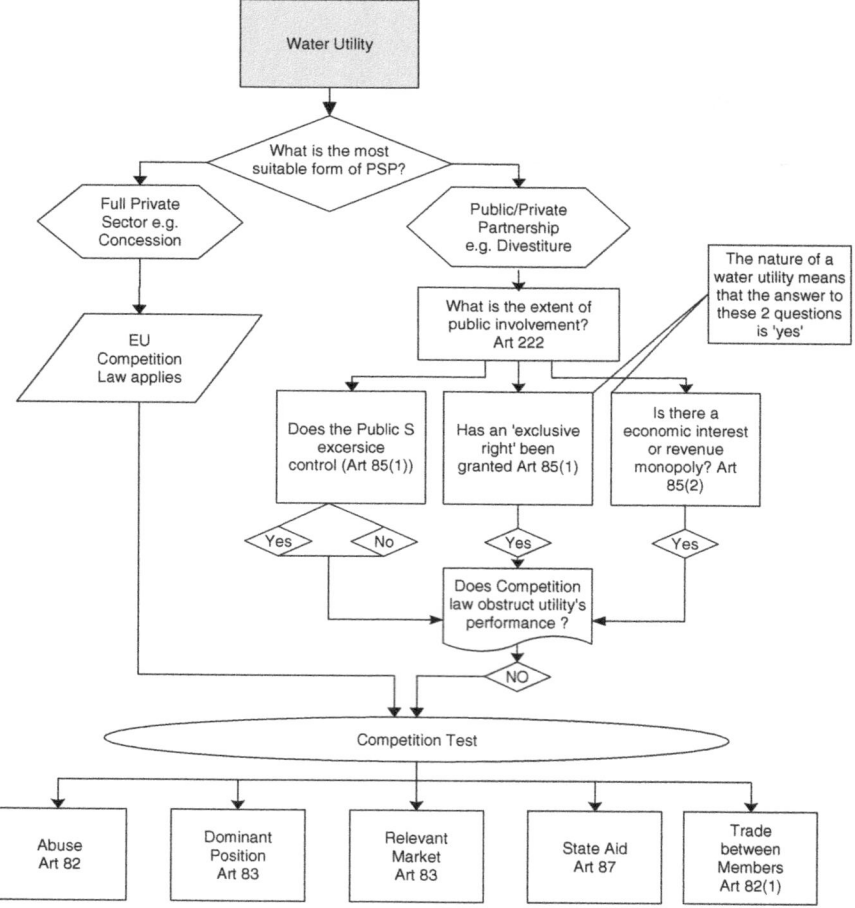

BIBLIOGRAPHY

A. PRIMARY SOURCES

A.1 Treaties

UN/ECE Convention on the Protection and Use of Transboundary Watercourses, Mar. 17, 1992, 31 I.L.M. 1312 (1992) (entered into force Oct. 6, 1996).
Treaty of Amsterdam amending the Treaty of European Union, the Treaties establishing the European Communities and certain related acts, 1999 O.J. (C340) 1 (entered into force May 1, 1999).
Convention on the Protection of the Rhine, 2000 O.J. (L289) 31 (not yet in force).
Treaty of Maastricht on the European Union, signed on 7 February 1992.
Treaty of Rome 1957.

A.2 EU directives

Directive 2000/60/EC of the European Parliament and of the Council of 23 October 2000 establishing a framework for Community action in the field of water policy, 2000 O.J. (L327) 1.
Council Directive 75/440/EEC on Surface Water for Drinking Directive, 1975 O.J. (L194) 26.
Council Directive 76/160/EEC on Bathing Water, 1976 O.J. (L31) 1.

© 2009 IWA Publishing. *Public and Private Participation in the Water and Wastewater Sector: Developing Sustainable Legal Mechanisms*, by X Cledan Mandri-Perrott. ISBN: 9781789065299. Published by IWA Publishing, London, UK.

Council Directive 76/464/EEC on Dangerous Substances, 1976 O.J. (L129) 23.
Council Directive 79/923/EEC on the Quality of Shellfish Water, 1979 O.J. (L281) 47.
Council Directive 80/68/EEC on Groundwater, 1980 O.J. (L20) 43.
Council Directive 91/271/EEC concerning Urban Waste Water Treatment, 1991 O.J. (L135) 40.
Council Directive 91/676/EEC concerning Nitrates, 1991 O.J. (L375) 1.

A.3 Other EU measures

Commission Proposal for a Council Directive establishing a framework for Community action in the field of water policy, COM (97) 49 final, 1997 O.J. (C184) 20.
Opinion of the Economic and Social Committee on European Water Policy, O.J. 1997 (C30) 5.
Opinion of the Economic and Social Committee on the 'Strengthening of the Law Governing Concessions and Public/Private Partnership Contracts', 2001 O.J. (C14) 91.
Communication from the Commission: The Challenge of Environmental Financing in the Candidate Countries, COM (2001) 304 Final, ELLIS Publications.
Parliament Report on the proposal for a Council Directive on establishing a framework for Community action in the field of water policy (COM (97) 49 final, COM (97) 614 final, COM (98) 76 final), 1998 O.J. (C343) 1.
Communication from the Commission to the Council and European Parliament, Establishing Pricing Policies for Enhancing the Sustainability of water resources, COM (2000) 477 final, Bulletin of the European Communities.
OECD Council Recommendation on Guiding Principles Concerning the International Economic Aspects of Environmental Policies, OECD Council Recommendation 1972 (C128) 72.

A.4 Legal documentation of PPP processes (Sofia and Tallinn)

Sofia Concession Agreement, (23 December 1999), Municipality of Sofia and Sofijska Voda AD concession agreement in relation to the supply of water and wastewater services for the Municipality of Sofia.
Pre-qualification Information Memorandum in relation to the Private Sector Participation to operate the Sofia water supply and wastewater system, currently operated by the municipally owned company, ViK Sofia (January 1999), Bulgaria.
Sofia Concession Agreement, (23 December 1999), Municipality of Sofia and Sofijska Voda AD concession agreement in relation to the supply of water and wastewater services for the Municipality of Sofia.
Information Memorandum, Tender Offer for Subscription of 30,000,000 A-Shares and Sale of 28,000,000 A-shares in AS Tallinna Vesi (Public Limited Company Incorporated Under the Laws of Estonia), (July 3 2000).
Resolution no. 210, City of Tallinn (June 15 2000), Estonia.
Shareholders Agreement, AS Tallinna Vesi, (Draft 24 November 2000), Estonia.

City of Tallinn and Tallinna Vesi. Services Agreement for (a) ensuring the functioning and maintenance of the public water supply and sewerage system of Tallinn, (b) extraction of fire fighting water from fire hydrants in the public water supply system and extraction of water from public water extraction points and (c) conduct of rain water, drainage water and other soil and surface water from public roads, streets and squares to the public sewerage system and treatment thereof. (Draft 24 November 2000), Estonia.

Tallinna Linn AS Tallinna Vesi and [Investor], Share Sale and Subscription Agreement, Final Draft 24 November 2000, Estonia.

Public Water Supply and Sewerage Act, February 10th 1999, (entered into force March 5, 1999).

Water Act, May 11 1994, Riigikogu Proceedings, TNN 1564 (1994).

B. SECONDARY SOURCES

B.1 Books

Blöch, H., The European Union Water Framework Directive: taking European water policy into the next millennium. (Water Science and Technology, 1999).

Bradbury, R., Privatization of Natural Monopoly Public Enterprises. The Regulation Issue, (Policy Research Working Paper 862. World Bank, Public Sector Management and Private Sector Development, Washington DC, 1992).

Cairns, W., Introduction to European Union Law, (Cavendish Publishing Ltd., London, 1997)

Cameron, J. & Werksman, J., Improving Compliance with International Environmental Law, (London, Earthscan Publications, 1996).

Castle, E.N., Berrens R.P. and Polasky, S., The Economics of Sustainability, 36 (NRJ 715, 1996).

Cook, P. and Kirkpatrick, C., Privatisation, Enterprise Development and Economic Reform, (Cheltenham, Edward Elgar Publishing, 1998).

Delmon, J., Privatisation, (London, Sweet & Maxwell, 2000).

Delmon, J., Water Projects: A commercial and Contractual Guide, (Kluwer Law International, Netherlands ISBN 90-411-1685-0, 2001).

Druesne, G., Droit matériel et politiques de la Communauté européenne, (PUF Paris, 1991) (in French).

European Commission Communication on Environmental Integration, A Strategy for Integrating Environment into EU Policies, (Journal of Environmental Law Vol. 11, Oxford Uni. Press 1999).

European Commission, The institutions and Bodies of the European Union: Who's who in the European Union? What difference with the Treaty of Nice make?, (Luxembourg: Office for the Official Publications of the European Communities, ISBN 92-894-0490-6, 2001).

Fontaine, P., Europe in Ten points, (European Commission, N° cat. PD-13-98-257-EN-C, 1998).

Hartley, T.C., The Foundations of the European Community, (Clarendon Press, London, 1994).

Howsam, P., (ed)., Water Policy (London, E & FN Spon Publishers, 1996).

Hobsbawm, E., Age of Extremes: The Short Twentieth Century, (Michael Joseph, 1994).
Idelovitch, E. and Ringskog, K., Private Sector Participation in Water Supply and Sanitation in Latin America. (Washington, DC: World Bank, 1995).
Kirby, J., O'Keefe, P. and Timberlake L., (eds.), Sustainable Development (London, Earthscan Publications 1996).
Lasok, D., Law and Institutions of the European Union, (Butterworths, London, 1994).
Maksymiuk, Zbigniew, Transformation de L'Enterprise Communale en Societe Joint-Venture: L'Example de la Societe Anonyme Saur Neptun Gdansk. (Gdansk. World Bank, Washington DC, 1994).
Munasinghe, M., Water Supply and Environmental Management: developing World Applications, (Boulder Colorado, Westview Press, 1992).
Nunes Correia, F., (ed)., Institutions for Water Resources Management in Europe, (Rotterdam, A.A. Balkema Publishers, 1998).
Rivera, D., Private Sector Participation in the Water Supply and Wastewater sector: Lessons from Six Developing Countries, (The International Bank for Reconstruction and Development, The World Bank, Washington DC, 1996).
Robinson, N.A., (ed.), Agenda 21 and the UNCED Proceedings (Oceana Publications, 1992).
Sands, P., Principles of International Environmental Law, (Manchester, Manchester Uni. Press, 1995).
Swann, D., Competition and Industrial Policy in the European Community (Methuen, 1983).
Vinter, G., Project Finance, (London, Sweet & Maxwell, 1998).
Whish, R.P., Competition Law, (4th Edition), (London, Butterworths, 1994).
Weigall, D., and Stirck, A., (Eds), The Origins and Development of the European Community, (Leicester University Press, England, 1992).
Yepes, G., and Dianderas, A., Performance Indicators. Water and Sanitation Utilities: Financial Indicators and Overview of Service Rates. (2nd Edition), (World Bank, Transport, Water, and Urban Development department, Washington, DC, 1996).
Vorkink, A.N., CASE STUDY: Bulgaria. (1999), Part I, 4, World Bank Publications, Country Director Bulgaria and Romania Country Unit, ECSPF.

B.2 Articles and presentations

Barlow, M., *Commodification of Water*, Proceedings of an International Conference, the 10th Stockholm Water Symposium, Stockholm, Sweden, (August 17, 2000).
Bryce, S., *The Privatisation of Water*, Nexus Magazine, Vol. 8, No. 3, (2001).
Bitran, E and Serra, P., *Regulation of Privatized Utilities: Lessons from the Chilean Experience*. OECD Advisory Group on Privatization, Eight Plenary Session, Paris, October 30 and 31. Paris. (1995).
Brook Cowen, P., *Private Sector Participation in the Water Sector: Designing an Appropriate Reform Option*, Proceedings of an International Water Seminar, June 8-11, 1998, University of Dundee, Scotland, (June 7–10, 1999).
Brook Cowen, P., *Ensuring Access to Water: Private Sector Participation and Services*, Proceedings of an International Water Seminar, University of Dundee, Scotland, (June 7–10, 1999).

Brook Cowen, P., *Competition and Regulation in Water and Sanitation*, Proceedings of an International Water Seminar, University of Dundee, Scotland, (June 8–11, 1998).
ENDS Environment Monthly, *Ofwat's Crushing Defeat over Environmental Investment'*, ENDS Report 289, (Feb. 1999).
European Bank for Reconstruction and Development, Sofia *Bulgaria Funding*, File name 438sofia.doc PSD Provided by the Bulgaria Project Desk of EBRD. (First issued 7 Sept. 2000).
Hemming, R., and Mansoor, A.M., *Privatization and Public Enterprises*, Internal Paper No 56, International Monetary Fund, Washington DC, (1988).
Johnstone, N., Wood, L. and Hearne, R., *Private Sector Participation in Urban Water and Sanitation*, 23 NRF 287 (1999).
Labre J., *The Role of the Private Sector*, Proceedings of an International Water Seminar, University of Dundee, Scotland, (June 7–10, 1999).
Lee, T.R., Alternatives for Private Participation in the Provision of Water Services, 20 NRF 333 (1996).
Legge, D., *The Water Industry in a Regulated Environment*, Journal of Environmental Law, Vol. 12, No. 1, (2000).
McCoy-Thompson, S., *Public-Private Partnerships*, 22 NRF 87 (1998).
Newsletter of Secretariat of the 3rd World Water Forum, No. 61, (January 21, 2002).
Roy, D., *How can Regulation be made Effective*, Proceedings of an ABD Conference, Geneva, Switzerland, (April 28, 1998).
Rieu-Clarke, A., *The EU Water framework Directive: A Model for sustainable use of water resources?*, University of Dundee, 2002; Dublin Statement 1992: Agenda 21, Chapter 18 *'Sustainable Use'* is defined as to ensure adequate supplies of water of good quality are maintained for present and future generations, (1992).
Rees, J.A., Regulation and Private Participation in the Water and Sanitation Sector, 22 NRF 95 (1998).
Stiggers, D., Eyre, B. and Graham, S., *The Trinidad Management Contract*, AWRA Paper (1998).
The Economist Country Report: *Bulgaria*, FT Publishing, London, (1997).
Triche, T, *Private Participation in the delivery of Guinea's Water Supply Services*, (1990), Policy Research Working Paper 477. World Bank, Transport, Water, and Urban Development Department, Washington DC: World Bank.
Yaron, G., The Final Frontier: A Working Paper on the Big 10 Global Water Corporations, Toronto, Polaris Institute Publishers, (2000).

B.3 Reports and periodicals

Booz Allen Hamilton & Severn Trent Water International, Draft Report on the Concession Monitoring Unit, Sofia Concession Contract, (July 2000).
Report of the World Commission on Environment and Development (The Brundtland Report), *Our Common Future*, 43 (1987).
European Community Progress Report, *Towards Sustainability* (Luxembourg, Official Publications of the EC, 1997).
European Commission, *Regular Report from the Commission on Bulgaria's progress towards Accession*, 9, Brussels: Office for Official Publications of the European Communities, 13.11.2001, SEC 1744, (2001).

European Commission, *Regular Report from the Commission on Estonia's progress towards Accession*, 35, Brussels, 13.11.2001: Office for Official Publications of the European Communities, SEC(2001) 1744, (2001).
Environmental Assessment Report No.1, *Sustainable water use in Europe*, (Copenhagen, EEA Publications, 1999).
Environmental Assessment Report No.2, *The Environment in the EU at the turn of the Century* (Copenhagen, EEA Publications, 1999).
Environmental Assessment Report No.6, *Environmental Signals 2000* (Copenhagen, EEA Publications 2000).
European Environment Agency Report, *Water Stress* (Copenhagen, EEA Publications 2000).
ENDS Environment Monthly, *Water Framework Directive Agreed* (ENDS Report 305, June 2000).
ENDS Environment Monthly, *DETR pushes ahead with economic instruments for water* (ENDS Report 304, May 2000).
European Commission Convergence Report, *The European Economy – Economic Policy Guidelines 2000* (No.70, 2000).

B.4 Interviews

Telephone interview with Dr Geoff Thorpe, Managing Director for International Water in Cochabamba, Bolivia (3 February 2003).
Interview with Alain Lucassol, Senior Water Engineer, Africa and Middle East Region, World Bank, (23 January 2003), Washington DC.
Interviews with PA Consulting Group's Private Sector Advisory Group in particular Mssrs David J Stiggers and Dr Kenneth Rubin over a period commencing from May 2000 to January 2002, Washington DC.
Interviews with Director of the Public Sector and Infrastructure Sector, Latin America Mr Abel Mejia during a series of interviews in June 2001 and subsequently during December 2002.
Interview with Mr Christopher Jennings, Senior Water Specialist of the Inter-American Development Bank (23 July 2001), Washington DC.
Telephone interview with Mt Les Stachow. Head of PPP In Eastern Europe for Booz Allen&Hamilton. Author of the Report on the Concession Monitoring Unit, Sofia Concession Contract, *supra* note 242, (15 August 2000).
Telephone interview with Mr David Carr, Project Manager for the Sofia Concession Project on behalf of International Water. (14 January 2002).
Telephone interview with Mr Andrei Delchev, on behalf of the Municipality of Sofia (4 March 2002).
Telephone interview with Mr Thomas Maier. Head of Eastern Europe for the European Bank for Reconstruction and development. (4 March 2001).
Telephone interview with Mr Tim Lowe, Business Development Manager for International Water and Project Manager responsible for the Sofia Concession. (12 February 2002).
Telephone interview with Mr Ivar Virkus, Deputy Mayor of the City of Tallinn (12 December 2001).

Telephone interview with Mr Chris Shugart, Senior Banker, European Bank for Reconstruction and Development (15 October 2001).
Telephone interview with Mr Tim Lowe, Project Manager for the Tallinn, Estonia on behalf of International Water (23 March 2003).
Telephone interview with the Chairman of the Management Board and Director of Development, Mr Kalle Tiiter (16 October 2001).
Telephone interview with Mr Priit Koit, Director of Suprema Securities, Merchant banker responsible for the sale process of AS Tallinna Vesi (1 Mar 2000).

B.5 EU cases and European Court of Justice decisions

Case 209–215, 218/78 *Heintz van Landewijck v Commission* (1980) ECR 3125 at 3250.
Continental can Case OJ 1972 L17 and Case 27/76 United Brands v Commission (1978) ECR 207.
Case 62/86, AKZO v Commission (1991) ECR I-3359.
Case 27/76, United Brands v Commission (1978) ECR 273.
ECJ Decision 82/861, OJ 1982 L 360/36.
Case 155/73, *Sacchi* (1974) ECR 409.
Cases 188-199/80, *France v Commission* (1982) ECR 2545.

B.6 Internet resources

ADB: Developing Best Practices for Private Sector Investment in Infrastructure, (visited Sept. 5, 2001). http://www.adb.org/documents/books/developing_best_practices/water_supply/default.asp
Alcazar, M., Abdala, L. and M. Shirley, 'The Buenos Aires Water Concession', (visited Sept. 10, 2001). http://www.econ.worldbank.org/view.php?topic=20&type=5&id=1065
Asian Development Bank Report, 'Private Sector Participation', (visited Sept. 10, 2001). http://www.adb.org/documents/books/private_sector_participation/default.asp
Byron, N., 'Economic Instruments and Taxation for Natural Resource Management' (visited Apr.8, 2001). http://www.environment.gov.au/psg/economics/byron.htm
Barlow, M., 'Private Drain on Public Water', (visited Sept. 7, 2001). http://www.theglobeandmail.com/hubs/national.html
Danish EPA, 'Economic Instruments in Environmental Protection in Denmark' (visited Apr.10, 2001).http://www.mst.dk/udgiv/publications/20000
Environment Commissioner Margot Wallstrom in, 'Water of the Europeans Integrating economic assessment in the decision making process', International Conference, Lille, Sept.13–14, 2000 (visited Mar. 26, 2001). http://www.health.fgov.be/WH13/krant/krantarch2000/kranttekstsept/000915r04eu.htm
Enlargement and Civil Society, Conference Proceedings 1999, taken from http://europa.eu.int/comm/enlargement/docs/index.htm
European Freshwater Programme, 'Actions for Internalising Sustainability' (visited Mar.26, 2001). http://www.wwffreshwater.org/initiatives/wfd
EEB Open Seminar on EU Water Policy under the WFD, 'Making the WFD Work', Brussels, Nov.3–4, 2000 (visited Mar. 27, 2001). http://www.eeb.org/activities/water/EEB%20Seminar%20WFD.

European Freshwater Programme, 'What role for water pricing' (visited Apr. 3, 2001).http://www.wwffreshwater.org/initiatives/wfd/integration

Environmental Issue Report No.19, *Sustainable water use in Europe* (visited Apr.5, 2001).http://www.reports.eea.eu.int/Environmental_Issues/No_19/en

Environmental Assessment Report No.2, 'Integration of the Economy and the Environment' (visited Apr. 5, 2001). http://www.reports.eea.eu.int/Environmental_Assessment_No_2/en.pdf

Ganiage, D., 'Governance Dimensions of Regulatory Frameworks for Private Sector Development', (visited Sept. 7, 2001). http://www.adb.org/Documents/Conference/Seminar_Governance_Frameworks.htm

Guide to the Approximation of EU Environmental Legislation, 'Overview of EU Environmental Legislation', (visited Mar. 29, 2001). http://www.europa.eu.int/comm/environment/guide/part2d.htm

Hall, D., 'Water Privatisation: Global Domination by a few', (visited Sept. 10, 2001). http://www.corporatewatch.org.uk/magazine/issue12/cw12w5.htm

International Environment Reporter, 'EU Parliament Invokes Conciliation Procedures for final text of WFD', (visited Apr. 6, 2001). http://www.subscript.bna.com/SAMPLES/ier.nsf

OECD Report, 'Economic Instruments for Pollution Control and Natural Resources Management: A Survey' (visited Apr. 10, 2001). http://www.oecd.org.env/docs/epocgeei9835.pdf

Presidency Conclusions Helsinki European Council (visited 10 and 11 December 1999). http://europa.eu.int/rapid/start/cgi/guesten.ksh?p_action.gettxt=gt&doc=DOC/99/16|0|RAPID&lg=EN

Presidency Conclusions Helsinki European Council 10 and 11 December 1999. http://europa.eu.int/comm/enlargement/intro/criteria.htm

Smith, T.T., 'Designing Enforceable Environmental Requirements in the EEC', (visited Mar. 29, 2001).http://www.inece.org/2ndVol2/Smith2.html

UNEP, 'The Use of Economic Instruments for Sustainable Development', Meeting on March 2001 (visited Mar. 28, 2001). http://www.unep.ch/etu/etp/events/recent/ber_econinst.pdf

Sacerdoti, G., 'Private Infrastructure for Development: Confronting Regulatory Risks', International Conference Proceedings, Sept. 8–10, 1999, Rome, Italy (visited Sept. 6, 2001). http://www.worldbank.org/riskconference/papers.htm

Spencer, R., 'Diverting the Cash Flow: Beyond Water Privatisation', (visited Sept. 8, 2001). http://www.corporatewatch.org.uk/magazine/issue12/cw12w5.htm

UNDP Working Paper, 'Public-Private Cooperation in the Delivery of Urban Infrastructure Services', (visited Sept. 5, 2001). http://www.undp.org/ppp/gln/publications/working1.htm

UN Bureau for Development Policy, 'Public-Private Partnerships for the Urban Environment', (visited Sept. 6, 2001). http://www.undp.org/ppp/gln/publications.default.htm

Van den Berg, C., 'Water Concessions', (visited Sept. 6, 2001) http://www.worldbank.org/html/fpd/notes/

World Water Council's, 'Investing for the Water Future', (visited Aug. 30, 2001). http://www.worldwatercouncil.org/vision/documents/chapter5.pdf

World Bank Tool-kits, 'Options For Regulating Water and Sanitation', (visited Sept. 7, 2001). http://www.worldbank.org/html/fpd/wstoolkits/Kit1/annex4.htm

Bibliography

World Bank Toolkits, 'Is the Regulatory Framework Conducive to Private Sector Participation', (visited Sept. 9, 2001). http://www.worldbank.org/html/fpd/wstoolkits/Kit1/kitone25.html

Wragge, S., 'Public-Private Partnerships', (visited Sept. 10, 2001). http://www.atse.org.au/publications/focus/focus-wragge.htm

World Water Forum, EU Water Framework Directive, (visited Mar. 26, 2001). http://www.wwffreshwater.org/initiatives/wfd.htm

IWA Publishing's authorised EU representative for General Product Safety Regulations is Diane D'Arras, 15 rue Duret, 75116 Paris, France, e-mail: safety@iwap.co.uk.

Printed and bound by CPI Group (UK) Ltd, Croydon, CR0 4YY
22/04/2026
02094793-0001